"US science is wid
viewed as the enem
explains how his N ...powered him to become a remarkable 'science diplomat' who works toward solving some of the world's
most intractable problems. And because science is a universal language,
scientists are ideally suited to create strong ties of personal trust and
understanding between nations. With institutional support, many
younger scientists can also contribute. Might this book help to catalyze
new, vigorous interactions between the future leaders of US science and
their colleagues across the globe?"

—Bruce Alberts, PhD; Chancellor's Professor of Science
and Education, Emeritus, UCSF; President Emeritus,
National Academy of Sciences

"A common bond of humanity in dealing with global challenges is brilliantly
illuminated in this book, including collaborations to address volcanic eruptions, infectious diseases, and nuclear crises. Agre provides intriguing details of scientists at work in all corners of the world who have contributed to
humanitarian diplomacy by bringing their science, practical solutions, and
keen observations to those in need. The stories told in this book carry a
wealth of information that provides fascinating reading."

—Rita Colwell, PhD; former Director, National Science
Foundation; founder, CosmosID

"Peter Agre offers a rare and compelling perspective on how scientists can
transcend politics to spark dialogue, build collaboration, and promote
peace. In a time of rising global tensions and eroding trust between nations, his accounts—from Cuba to Iran to North Korea—reveal how scientists have helped shape pivotal moments in international affairs. These
stories remind us that scientific and political partnerships can create unlikely openings where traditional diplomacy stalls. Agre makes a powerful
case for science as both a tool of progress and a force for connection in an
increasingly fractured world."

—Victor J. Dzau, MD; President, National Academy of Medicine;
James B. Duke Professor of Medicine, Duke University

"Science diplomacy is one of our least shaped aspects of U.S. foreign policy, even though it could one day have the greatest impact. Over the last century, America's research universities and institutes became the envy of the world, and many of the most important leaders of nations studied in our country. Now, renowned Nobel Laureate Dr. Peter Agre and science communicator Dr. Seema Yasmin team up to offer new paths for leveraging our scientific achievements into essential and productive tools of diplomacy. They provide important lessons for future international relations, especially in a time when technological superiority has become one of the best predictors of regional and global dominance."

—Peter Hotez, MD, PhD; Professor and Dean, National School of
Tropical Medicine, Baylor College of Medicine; former U.S. Science Envoy

"In an increasingly multipolar world, the tenuous ties that bind are more vital than ever; science diplomacy is a tool that can hold us together when everything else is pulling us apart. In this essential narrative, Drs. Agre and Yasmin show the way forward for those willing to work at the intersection of science and international relations."

—Sudip Parikh, PhD; CEO, American Association for the Advancement
of Science (AAAS); Executive Publisher, *Science*;
member, Council on Foreign Relations

"Disease and natural disasters do not respect national boundaries, but Peter Agre has gone to places and held conversations that required scientific diplomacy where political diplomacy simply could not reach, offering better health and safer lives to people around the world. Today, we need more Peter Agres."

—Lois Quam; member, Council on Foreign Relations; board member,
Commonwealth Fund; former CEO, Pathfinder International

"Peter Agre has been where most politicians fear to tread. His book shows how science transcends geographical and political boundaries and creates relationships that last. All modern governments need science and technology; these human stories reveal the power of science as a diplomatic force."

—Lord Patrick Vallance, MBBS, UK Minister of State for
Science, Research, and Innovation

Can Scientists Succeed Where Politicians Fail?

JOHNS HOPKINS
WAVELENGTHS

In classrooms, field stations, and laboratories in Baltimore and around the world, Johns Hopkins University researchers explore the world's most complex challenges and vital opportunities. The Johns Hopkins Wavelengths program brings audiences inside their stories, presenting pioneering discoveries and innovations that benefit people in their neighborhoods and across the globe in artificial intelligence, bioastronautics, cancer research, epidemiology, food systems, health equity, marine robotics, planetary science, science diplomacy, and other critical areas of study. Through these compelling narratives, their insights spark conversations in dorm rooms, dining rooms, boardrooms, and the offices of leading government representatives—including the Oval Office.

This media program—which includes narrative nonfiction books, secondary school lesson plans, digital health toolkits, and exhibits—is a partnership between the Johns Hopkins University Press and the University's Office of the Vice Provost for Research. Team members include:

Consultant Editor: Seema Yasmin

Wavelengths Series Creative Director: Anna Marlis Burgard

Senior Acquisitions Editor: Matthew McAdam

Production Editorial Manager: Jennifer D'Urso

Senior Production Editor: Charles Dibble

Wavelengths Series Designer: Matthew Cole

Art Director: Molly Seamans

Production Manager: Jennifer Paulson

Publicists: Kait Howard, Core Four Media

JHUP Executive Director: Barbara Kline Pope

Associate Vice Provost for Research: Nicholas Wigginton

Can Scientists Succeed Where Politicians Fail?

PETER AGRE, MD

with Seema Yasmin, MB BChir

Johns Hopkins University Press
Baltimore

© 2025 Peter Agre
Cover illustration © 2025 Robert Hunt
All rights reserved. Published 2025
Printed in the United States of America on acid-free paper

9 8 7 6 5 4 3 2 1

Johns Hopkins University Press
2715 North Charles Street
Baltimore, Maryland 21218
www.press.jhu.edu

Cataloging-in-Publication Data is available from the Library of Congress.
A catalog record for this book is available from the British Library.

ISBN: 978-1-4214-5299-9 (paperback)
ISBN: 978-1-4214-5300-2 (ebook)
ISBN: 978-1-4214-5301-9 (ebook, open access)

The Open Access edition of this book is licensed under a Creative Commons
Attribution-NonCommercial-No Derivatives 4.0 International License: https://
creativecommons.org/licenses/by-nc-nd/4.0/. CC BY-NC-ND

*Special discounts are available for bulk purchases of this book. For more information, please
contact Special Sales at specialsales@jh.edu.*

EU GPSR Authorized Representative
LOGOS EUROPE, 9 rue Nicolas Poussin
17000, La Rochelle, France
E-mail: Contact@logoseurope.eu

Illustrating Science Diplomacy

Science diplomacy is an ever more urgent topic, as the recent pandemic and other devastating public health crises have taught us. It's also a fascinating field, one that yields the fruits of collaboration on many fronts. But it's also a challenging concept to portray, if the artist is to resist tired visual tropes.

Robert Hunt is an award-winning artist best known for the image he created of a boy (his son) fishing from a crescent moon for Steven Spielberg's DreamWorks Pictures. A veteran illustrator for *The Atlantic, Rolling Stone, Scientific American,* and *The Wall Street Journal* as well as a book jacket illustrator for hundreds of editions, he is no stranger to complex subject matter. Whether in his Bay Area studio, along the Tour de France route, or in Nepal as he prepared to ascend Mount Everest, Hunt creates beautiful, intelligent, and memorable works. On the topic of this book's cover image, he wrote:

"I was thrilled and honored to be asked to paint the cover art for this important book. The final work was the product of a collaboration between myself and the Wavelengths Creative Director Anna Marlis Burgard; the point of origin was a desire to respect and acknowledge the material, the author, and the reader. There were many rounds of conceptual sketches before the direction for the cover was agreed upon. Anna proposed including stamps from Dr. Agre's passports (those from North Korea as well as Zambia), which led me to a related idea: incorporating elements of the book's content—points on the globe, malaria cells, volcanic eruptions—into spheres as well. The arrangement of the spheres represents the cyclical evolution of science and progress, and the structure of the atom and the solar system. The clockwork gears are intended to represent the interconnectedness of everything, through the passage of time. The final image is a 24″ × 24″ oil painting."

Contents

Preface

SCIENCE IN ACTION

ON A DECEMBER AFTERNOON IN 2011, eight middle-aged men shared a meal of fried fish and beer in Havana's Old Town while they chatted enthusiastically about science. Among them were two Cuban climate scientists, the Cuban Academy of Sciences' director, and a Cuban nuclear physicist—who also happened to be the son of the former president, Fidel Castro. I was one of the four American scientists. I sat next to Dr. Fidel Castro Díaz–Balart, also known as Fidelito ("little Fidel"), a soft-spoken fellow who otherwise resembled his father. We were all delighted that our gathering of Cuban and US scientists was now possible.

Cuba had been closed to Americans since Castro established a socialist state in 1959 when I was 10 years old.[1] The October 1962 Cuban Missile Crisis brought the world to the brink of thermonuclear war.[2] Embargoes and sanctions followed, blocking the free flow of ideas between scientists. Opportunities had been lost, and certainly some scientific discoveries lay uncharted because politicians had prohibited scientists from working together and advancing their fields.

Decades later, a new American president had eased, though not fully reversed, those restrictions, making our trip possible.

President Barack Obama used executive powers to repeal restraints.[3] So, when the American Association for the Advancement of Science asked me to visit Cuba as part of its Center for Science Diplomacy, I promptly said yes—because I could.

I had dreamed of such a visit since I was a youngster in Minnesota, tinkering in my father's college chemistry laboratory a short stroll across campus from our house. The long winter nights made it easy to conjure images of the exotic, far-flung places where science could take me. Cuba was especially fascinating—a tropical island run by young, bearded revolutionaries who played baseball in their military fatigues.

The Havana lunch would have been difficult a few years earlier, and the conversation with Fidelito might not have occurred at all. Instead, midway through our meal, Fidelito leaned over to pass on a message. "Fidel Castro wants to meet you," he whispered. Despite my regimented schedule, his father had a plan, and he asked that I wrap up my afternoon lecture early so we could talk. I sipped more beer to hide my smile.

We rode with Fidelito in his spotless white Lada along a boulevard lined with beautiful villas and tropical gardens. We met in a back room of a conference center where I witnessed firsthand that the aged revolutionary had not lost his fierce spirit. The conversation with Fidel Castro lasted more than three and a half hours. Not surprisingly, Castro did most of the talking. Instead of tiring, he seemed to gain energy as the hours passed. Professor Alan Robock from Rutgers University and I were the only foreign guests. At 85 years old, Castro was slightly stooped and a little unsteady when standing, but his mind was sharp. He had bright eyes and spoke with a raised fist. He jabbed the air with his index finger to emphasize his words. With his son, Antonio, an orthopedic surgeon, and Fidelito on one side, and his wife on the other, the garrulous elder statesman ranked American presidents in order of untrustworthiness. He opined about US foreign policy, and he regaled us with details of new anticancer vaccines developed by Cuban scientists that were yet to be shared with the outside world. Castro spoke with pride about the Escuela Latinoamericana de Medicina, also known as the Latin American School of Medicine, which he had founded, and the hundreds of Cuban-trained doctors and nurses who provide medical aid to the people of Venezuela, Guinea-Bissau, the Congo, and Angola.[4] Despite Castro's admiration for science and his passion for universal health care, he demonstrated no remorse for Cuba's human rights violations and moribund economy.

In Cuba, as in many parts of the world, politics had derailed research and made collaborations between scientists impossible. And yet, as a child, I had seen how politics could spur scientific innovation. When the Soviet Union launched Sputnik, the first satellite to orbit Earth, in October 1957, I watched the adults on TV whisper their concerns about the need for the US to catch up in the space race. The next month, the Soviets launched Sputnik II, which took Laika, a stray dog from the streets of Moscow, into orbit, making her the first living being to visit space.

The US responded with vigor. Concerned for national security, the US government offered strong bipartisan support to increase spending on science, technology, engineering, and math. President Eisenhower signed bills to create the National Aeronautics and Space Administration (NASA), the National Defense Education Act (NDEA), and the Defense Advanced Research Projects Agency (DARPA).[5] International rivalry fueled scientific advancement. Increased funding for sabbatical fellowships from the National Science Foundation meant that our family could spend a year in the Bay Area while my father pursued research in the chemistry department at the University of California, Berkeley.

The American people lapped up the sudden passion for scientific progress. I was glued to the TV when Glenn Seaborg, the Nobel Prize–winning chemist, used mouse traps and ping pong balls to explain a chemical chain reaction and when NASA director Wernher von Braun demonstrated rocket science. Even America's first great theme park, Disneyland, joined in the fun. In 1967, a massive overhaul of the theme park launched *Tomorrowland,* an exhibition of futuristic lives offering a glimpse into a new tomorrow, one brimming with house-cleaning robots and vacations on the moon.

If political rivalry could spur a new space agency, put science on TV, and simulate space launches in Disneyland, couldn't political cohesion fuel even greater innovation? Good science demands collaboration, and collaboration relies on political harmony. In the twentieth and twenty-first centuries, global war and peace have hinged on scientists working alongside political negotiators: The Paris Climate Agreement, the Iran Nuclear Accord, the Manhattan Project, the Antarctic Treaty of 1959, the Partial Test Ban Treaty of 1963, and the Strategic Arms Limitation Treaty of 1972 are just some of the international cooperative agreements that have required the work of American scientists alongside policymakers. Sometimes in secret, sometimes as official ambassadors for their governments, scientists have swapped out white coats for blazers, stepped out of the lab and into US Capitol buildings, and played critical roles in political negotiations.

Some trace the origins of this scientific-political intermingling and the emergence of scientist-activists to 1945 and the unleashing of American-made atomic bombs on Japan. The American physicist J. Robert Oppenheimer, wartime director of the Los Alamos Laboratory and so-called father of the atomic bomb, led a group of Manhattan Project scientists to lobby the US government for international treaties to curb the use of atomic weapons. Oppenheimer continued his peace advocacy after the war as advisor to the US Atomic Energy Commission, pushing for international control of nuclear proliferation and voting against the development of the hydrogen bomb in 1949.[6]

The next year, physicist and engineer Lloyd Viel Berkner published a report advising the US State Department to weave science into its daily operations and appoint scientists in key roles.[7] Over the next two decades, US policymakers created formal positions for scientists as technical advisors and attachés, deploying some even inside the White House. But the mixing of science and politics and the leveraging of scientific knowledge for peace and political gain dates back even earlier. When the American colonies needed France's help to defeat the British, they sent Benjamin Franklin as minister, not only because Franklin had connections to the French government but because he was a scientist and could represent the colonies as an enlightened and forward-thinking emerging nation.[8]

My own family's immigration from Scandinavia to the United States in the 1870s was made possible by politicians who sought European workers to farm the land. Weeks after President Abraham Lincoln signed the Homestead Act of 1862, he signed the Morrill Land Grant Act. The Homestead Act encouraged settlement of the western territory, while the Morrill Land Grant Act created new universities with a focus on agricultural and mechanical subjects, with the aim of providing advanced education to the children of new arrivals. The next year, at the height of the Civil War, Lincoln signed a bill to create the National Academy of Sciences at the urging of a group of Massachusetts scientists.

My first taste of science merging and clashing with politics occurred when I was 15 years old. An eccentric, two-time Nobel prize–winning chemist stayed at our home. Linus Pauling received the 1954 Nobel Prize in Chemistry for studies of the nature of the chemical bond, which led to three-dimensional structures of complex molecules. Pauling also identified the cause of sickle-cell anemia—the first known molecular disease caused by a genetic defect. The discovery of sickle hemoglobin was made with the help of Harvey Itano, a young Japanese-American biochemist and physician who had been incarcerated in a World War II internment camp[9]. Pauling's Caltech peers ostracized him for working with a scientist of Japanese descent. Itano eventually left Pauling's lab to work with James V. Neel, with whom he discovered other hemoglobin variants.

Although Pauling had made major contributions to chemistry, biology, and medicine, it was his moral compass and activism that earned him the 1962 Nobel Peace Prize. As an ardent and vocal opponent of nuclear weapons testing, Pauling spurred President Kennedy to aggressively push for the limited test ban treaty. Pauling's extensive travels to lecture on the dangers of nuclear weapons aroused suspicions in the State Department, which ultimately seized his passport. Government officials suspected him of harboring communist sympathies, an accusation that he vigorously denied.

My father regarded Pauling as a hero, and while I was only a teenager, I was glad to agree. My father, chairman of the chemistry department at Augsburg College (now University) in Minnesota, invited Pauling to visit. Pauling gave a talk about chemistry in the morning and a public lecture on world peace in the evening. And while I respected Pauling's prizes and contributions to science, it was his work outside of the lab and the way he used his credibility as an accomplished scientist to lobby politicians that was uniquely inspirational. Despite being an internationally prominent scientist, he was a warm-hearted gentleman. He shared our family meals, and he graciously attended our evening devotions, making my mother (a farm girl who never attended college) and the rest of us aware of his appreciation.

Meeting the esteemed scientist-activist helped me realize that while I was fascinated by science, I was equally enthralled by the people who made fundamental discoveries. I was even more impressed by the impacts they had on geopolitics and culture. Pauling inspired me to imagine a scientific career that would include laboratory discoveries but could propel me from behind the bench and into every corner of the world. I envisioned science as an access pass into nations that were closed to Americans. I was far from certain that I had the talent needed to succeed in laboratory science, but I still hoped for a career that put science into action.

Two years after Pauling's visit, a high school camping tour of Europe and the Soviet Union consolidated that ambition. It was 1966, the height of the Cold War and the escalation phase for the United States in Vietnam, a period when the ethical quandary of scientists working on weapons research was again brought to the fore. It was eye-opening to observe and compare differences in geography, architecture, culture, and political systems in Scandinavia, Russia, Ukraine, the Caucasus, and the Balkans, but the friendly curiosity the people always had for us teenagers was universal.

I began to not just imagine but plan a career that would make medical science my global diplomatic pass. I soon realized that the path would be long. First, I graduated from the Johns Hopkins School of Medicine; then, after a decade of postdoctoral study in clinical and basic sciences, I was appointed assistant professor of medicine and assistant professor of cell biology at Johns Hopkins in 1984. But I did as my teenaged self had hoped: After decades of study and research, I established a laboratory in which my team investigated human red blood cells and advanced the field in a significant way. We discovered the first known membrane water channel, called aquaporin-1, which selectively facilitates the entry or release of water from cells. That discovery earned me the Nobel Prize in Chemistry in 2003.

Receiving a Nobel Prize undeniably changes a scientist's life. With a major increase in visibility, my lab received countless invitations from the global science community. And while the fanfare that accompanies the Nobel Prize was not my intention, the prize opened the door to the vision that Pauling had helped to conjure all those years ago: the Nobel Prize was the diplomatic pass I had dreamed of. Even after two decades, interest

Aquaporins and the Nobel Prize

In 1992, our team at Johns Hopkins University discovered the first molecular water channel, aquaporin-1 (AQP1). This protein explains how the plasma membrane of certain cells, such as red blood cells and kidney tubules, is so rapidly permeated by water.

Twelve other related human aquaporin genes were subsequently identified—each specialized to facilitate water homeostasis in different tissues. The term "aquaporin" refers to the entire family of water channels, which are known to exist in all forms of life (mammals, fish, insects, parasites, plants, fungi, and microbes). Aquaporins generate the aqueous component of cerebrospinal fluid, aqueous humor, tears, saliva, bile, airway secretions, amniotic fluid, sweat, and urine in humans, as well as rootlet water uptake, stem turgor, and transpiration in plants.

Some aquaporin functions have long been recognized. One example is Ivan Pavlov's dog salivation experiments, for which he won the 1904 Nobel Prize in Physiology or Medicine. Aquaporins play pivotal roles in diseases such as heart failure, kidney failure, brain swelling, cataracts, and even malaria and other parasitic diseases.

Our studies over two decades included collaborations with scientists from the United States, Denmark, Norway, Netherlands, Switzerland, Germany, France, India, Japan, China, South Korea, and a dozen other countries. Hundreds of lecture trips allowed me to personally spread the word about aquaporins. The response from the global scientific community was enormous, and multiple scientific groups worldwide joined this important quest. For this discovery, I shared the 2003 Nobel Prize in Chemistry with Roderick MacKinnon of Rockefeller University, who discovered the structure of potassium channels.

FURTHER READING

Preston, G. M., T. P. Carroll, W. B. Guggino, and P. Agre. "Appearance of Water Channels in Xenopus Oocytes Expressing Red Cell CHIP28 Protein." *Science* 256, no. 5055 (1992): 385–387. https://doi.org/10.1126/science.256.5055.385.

in the prize has not diminished, and the prize continues to provide opportunities to observe and participate in global events that are the basis of this book.

I have had the chance to meet with scientists in countries led by autocratic regimes hostile to the United States. Not surprisingly, it is generally impossible for agents of the US government to otherwise visit countries such as North Korea and Iran. But while serving as president and chair of the board of directors of the American Association for the Advancement of Science (AAAS) from 2009 to 2011, I participated in its new science diplomacy program by leading visits of nongovernment US scientists to Cuba, Iran, and North Korea. The groups I visited with were able to foster friendly relations with individual scientist-leaders in all three countries.

Ten years after discovering aquaporin-1, my lab shifted our focus from water channels inside cells to the parasite that infects red blood cells, causing malaria in a quarter of a billion people annually. As director of the Johns Hopkins Malaria Research Institute at the Bloomberg School of Public Health from 2008 to 2023, I oversaw and supported malaria field studies in Zambia, Zimbabwe, and the Democratic Republic of the Congo. In Zambia, I befriended an American pediatrician and an Oxford-educated malaria scientist leading heroic efforts to eliminate malaria.

These stories of innovation and advancement sit alongside darker chapters. A global diplomatic passport not only opens the door to collaboration; it provides close encounters with the human rights atrocities experienced by some scientists working in authoritarian states—and it entails a duty to defend their right to conduct scientific work. Advocating for the release of imprisoned scientists in countries such as Libya became the crux of my involvement as chair of the Committee on Human Rights at the National Academies of Sciences, Engineering, and Medicine from 2005 to 2008.

Taken together, these stories illustrate the theme that scientists have special opportunities to contribute, directly or indirectly, to improving US international relations with countries in which politicians have failed to build bridges. Contributing to our national interest by serving as ambassadors for both science and diplomacy, scientists have unique and invaluable opportunities to deescalate tensions and long-standing animosities between countries. Offering guidance on how to build scientific partnerships can forge a pathway to peace and prosperity.

Through trips to Iran, North Korea, Cuba, sub-Saharan Africa, and beyond, science became my conduit for bridge-building with scientists in nations considered hostile or even enemy states. More than mere knowledge exchanges, these visits led to meetings with ambassadors and presidents, possibly enhancing the machinery of political negotiations far removed from my laboratory research. I haven't been alone in this journey. Hundreds of American scientists crisscross the planet each year, shaking hands with peers, speaking to students, and presenting at conferences. This book explores how relationships forged across borders can facilitate international political negotiations, such as the nuclear agreement with Iran.

Malaria deaths continued to rise during the COVID-19 pandemic. The challenges facing humanity—including terrorism, infectious diseases, and climate change—continue to increase in scope and scale. Science will provide technical solutions to some of these existential threats. But scientists can bring much more to the diplomatic table: Where politicians fail to reach consensus, scientists push for peace, ease stalemate negotiations, and remind us that challenges connect humanity more than they divide.

CHAPTER 1

Cuba

SCIENCE UNDER SANCTION

HOW COULD SCIENCE SUCCEED IN CUBA when internet access was spotty? Arriving in Havana for the first time in November 2009, I struggled to get a solid broadband connection, even at Havana's Hotel Nacional de Cuba. It was a temporary inconvenience for me, one that would last as long as my short trip. But for 11 million Cubans, the economic damage from a half-century of trade embargoes meant chronic deficiencies of food, housing, transportation, sanitation, water—and functional internet access.[1] These were compounded by human rights abuses from a government led by a strong-willed revolutionary who remained in power concurrent with the terms of 11 US presidents. A living artifact of the Cold War, Cuba somehow not only survived but excelled in some areas of science despite its spotty internet.[2]

I had been eager to visit Cuba for decades. As the incoming president of the American Association for the Advancement of Science (AAAS) in 2009, I knew it was important to build bridges between my fellow US scientists and our colleagues in

Cuba. A year earlier, AAAS had launched its Center for Science Diplomacy led by Vaughan Turekian and Norman Neureiter.[3] Among the center's primary objectives was promoting science as a tool to establish relations with hostile states. Cuba was of particular interest. There was exciting news about the nation's significant advancements in cancer treatments and medicines for eye disease. Still, despite a promising visit to Cuba by National Science Foundation Director and AAAS President Rita Colwell in 1997, US relations with Cuba remained caustic. Trade and travel embargoes designed to cripple the Cuban economy and destabilize the Castro regime had devastated scientific relations.

The plan was for a small delegation of US scientists to travel to Cuba to confer directly with Cuban scientists.[4] This would allow extended meetings with the leadership of the Cuban Academy of Sciences and introductions to medical scientists to learn about Cuba's development of new medical interventions: an injectable treatment for diabetic foot ulcers, low-cost vaccines, and cancer treatments—including a therapeutic vaccine to treat lung cancer, a disease-screening program for newborns, and educating hundreds of Cuban-trained doctors sent each year to provide medical assistance in poorer countries.[5]

As we will see in this chapter, all this occurred despite lingering tensions over a scientific snub in the early 1900s and a 50-year-long trade embargo that made scientific equipment and reagents inaccessible.

FIDELITO

The invitation came from the eldest son of Fidel Castro, Fidel Castro Díaz-Balart, who became our colleague and friend. Bearded and physically resembling a shorter version of his father, he was known affectionately as Fidelito, or "Little Fidel." Unlike his father, Fidelito was soft-spoken and reserved. After receiving his PhD in nuclear physics from Moscow State University, Fidelito returned to Cuba, where he was made head of the Cuban atomic energy agency. Fidelito directed the building of a nuclear power plant station at Juragua near Cienfuegos, Cuba. The construction was problematic due to equipment and material shortages and inefficiency. After the collapse of the Soviet Union in 1991, the project was finally abandoned, and Fidelito was publicly fired from this position by his father. Fidelito slowly recovered his stature and eventually became vice president of the Cuban Academy of Sciences and senior science advisor to the president of Cuba—initially his father and subsequently his uncle Raúl.[6]

With Fidelito's introductions, I made seven trips to Havana between 2009 and 2017 to visit the Cuban Academy of Sciences, biotech institutes and laboratories, universities, teaching hospitals, and government ministries. Most of the senior scientific leaders in Cuba were elderly and had lived through the revolution. These veteran scientists exhibited a patriotic fervor for their hero, whom they usually referred to by his first name—

Fidel. While none of the scientists were critical of the American people, most expressed deep resentment over the US government's trade embargoes.

Most of Cuba's current researchers have only known science under embargo. While some of the older generation exhibited a pugnacious character reflective of their revolutionary hero, those born after the revolution seemed relatively uninterested in Fidel Castro and the Cuban Communist Party. It was clear that these young scientists, to whom Fidelito was looking for Cuba's future, would become increasingly important. Building bridges with them was critical. We introduced Fidelito to the organizers of the summer Lindau Meetings of the Nobel Laureates, an annual gathering in Bavaria. Fidelito and three Cuban graduate students in physics attended the meeting in 2012, where they spent one week in residence with 27 Nobel Laureates in Physics and more than five hundred of the world's best graduate students from 69 countries. They were the first Cubans ever to participate in the Lindau Meeting.

A scientific snub may have fueled some of Cuba's advances and discoveries (more on this later), but so did its drive to make scientific strides despite its circumstances. While Cuba is the largest island in the Caribbean, its total gross domestic product is around half of the $140 billion the United States invests in research annually.[7] Through necessary ingenuity and the prioritization of scientific advancement under former President Fidel Castro, Cuba developed a cluster of biotechnology research companies to rival some in the West—even as Cuban scientists have been mostly unable to travel to the United States to share their research.

THE CUBAN REVOLUTION AND SANCTIONS

US trade restrictions on Cuba date back to the Cuban Revolution, which ended in 1959, when Fidel Castro and his revolutionaries ousted the US-backed president, Fulgencio Batista, who had taken power in 1952 in a bloody coup that left more than twenty thousand Cubans dead.[8]

Trade restrictions have tightened over the decades. President Dwight Eisenhower's administration recognized Castro as Cuba's new president but banned almost all US exports to Cuba, slashed Cuban sugar imports, and imposed further economic penalties as Castro forged stronger ties with the Soviet Union.

Two years after the revolution, the situation worsened. Only 90 days after his inauguration, President John F. Kennedy's administration staged the disastrous Bay of Pigs Invasion of Cuba.[9] The botched CIA attempt to oust the Cuban leader led to the Soviets' installation of nuclear weapons in Cuba, with ballistic missiles stationed just 90 miles from American shores. The missiles were spotted in October 1962 when the pilot of an American U-2 spy plane flying over Cuba photographed a Soviet SS-4 medium-range missile being assembled. Those photographs set off the Cuban Missile Crisis, a 13-day standoff between the Soviet Union and the United States that threatened to turn the Cold War into a thermonuclear Armageddon.[10] A full economic embargo followed, including the strict travel restrictions that have impeded science ever since.

The freeze persisted since 1959 but began to thaw in 2009 when the newly elected US president, Barack Obama, loosened travel restrictions to Cuba. Then, in December 2014, Obama abruptly changed US policy when he announced, during a televised speech, that he would "cut loose the shackles of the past" and finally end the "rigid policy that is rooted in events that took place before most of us were born."[11] Obama announced the restoration of full diplomatic relations with Cuba, including an end to the stringent travel restrictions, and became the first sitting president to visit Cuba in 88 years.[12] American scientists, educators, and artists with reasons to travel to Cuba had been able to visit the island with permission in prior years, but in 2016, JetBlue Flight 387, the first commercial flight to Cuba from the United States since 1961, departed Fort Lauderdale, Florida. Cuba was finally opening up to the United States.

YELLOW FEVER

Obama's executive actions still prevented Americans from visiting Cuba solely as tourists, instead necessitating a reason sanctioned by the US Treasury, including business, religious, and educational travel or a family visit. But in reversing US policy, the president referenced a point of science history: "It was a Cuban, Carlos Finlay, who discovered that mosquitoes carry yellow fever; his work helped Walter Reed fight it," the president said in his 2014 speech. "We've seen the benefits of cooperation between our countries before."[13]

Obama referenced one of the landmark cases of international scientific collaboration. A Cuban doctor and epidemiologist, Carlos Juan Finlay, discovered how yellow fever spread. The discovery revolutionized the management of yellow fever, a deadly disease that caused periodic epidemics in Cuba and the United States and shaped the course of history.[14]

Cuban doctors had long been puzzled by Havana's yellow fever outbreaks. Sometimes thousands fell sick with crippling pains and fevers, and hundreds would die. But why was Havana—a city with clean streets and spacious quarters—the epicenter of yellow fever outbreaks in the late 1800s? The same scourge had killed more than twenty thousand Americans in 1878 alone, but they were mostly poor immigrants living in squalor in the southeastern United States. How could yellow fever spread through Havana so rapidly if the virus was transmitted through dirty clothing, bedding, and close contact with the sick?

The US Army had a vested interest in solving the yellow fever mystery. But it was mystified as to how it spread among its troops, often killing even the young, healthy, and strong. In 1900, scientific knowledge rapidly expanded, and the ability to distinguish small parasites from bacteria was possible; viruses remained enigmatic and were often referred to as "filterable agents." They weren't visualized until the invention of the electron microscope in 1931. Assuming that infection spread through the air, pathologist and Army Major Walter Reed told a friend that a "plug of cotton in the nostrils would be advisable" while stationed in Cuba. Major William Crawford Gorgas, the chief sanitary officer in Havana, postulated that the infection was spread "by filth, dirt, and general unsanitary conditions."[15] But after a cleanup that diminished the spread of dysentery and typhoid in Havana, yellow fever attacked again and again. "The health authorities were at their wit's end," wrote Gorgas. "We evidently could not get rid of Havana as focus of infection by any method."

Scientific progress is delayed when errors are left unchallenged by other scientists. Such a situation arose in the history of yellow fever. The July 1893 issue of the *British Medical Journal* published a report by an Italian bacteriologist working in Montevideo, Uruguay. Giuseppe Sanarelli, on the basis of isolates taken from yellow fever patients, proposed that a bacterium, *Bacillus icteroides*, caused yellow fever.[16] European scientific leaders embraced the idea. But US Army Surgeon General George Sternberg, who had pursued the cause of yellow fever for years, was convinced that Sanarelli was wrong.

Compelled to refute Sanarelli, Sternberg appointed Walter Reed to organize the Yellow Fever Commission to Cuba. Aware of the work of Carlos Finlay, Reed suggested that they study the role of the mosquito. Sternberg replied, "No; that has already been decided to be a useless investigation." Two decades earlier, Carlos Juan Finlay, the Cuban physician and epidemiologist referenced in Obama's 2014 speech, proposed a theory of transmission of yellow fever. The Cuban-born doctor of Scottish and French parents graduated from Jefferson Medical College in Philadelphia in 1855 and turned down a lucrative career in the United States to practice in Cuba. While yellow fever consumed most of his passion and time, Finlay studied other infectious diseases as well, such as cholera and tetanus. His theory about the cause of an 1867 cholera outbreak—that the public water supply had been contaminated by sewage—was met with disbelief. That didn't stop Finlay from presenting scientifically sound theories.[17]

Finlay had been dubbed a "crazy old man" and a "crank" by those who doubted his theory of mosquitoes as yellow fever vectors. On August 14, 1881, he presented a paper to the Cuban Academy of Sciences titled "The Mosquito Hypothetically Considered as the Transmitting Agent of Yellow Fever." Finlay knew that evidence to support his hypothesis was weak, but he didn't anticipate the ridicule that followed. "I understand but too well that nothing less than an absolutely incontrovertible demonstration will be required before the generality of my colleagues accept a theory so entirely at variance with the ideas which have until now prevailed about yellow fever," he said.[18]

INTERNATIONAL COOPERATION AND THE MOSQUITO HYPOTHESIS

For the next two decades, the Cuban doctor conducted more than a hundred experiments with mosquitoes and humans in attempts to prove his hypothesis. The evidence never stacked in his favor. In 1884, Finlay's nascent hypothesis about the spread of yellow fever via mosquitoes was published, albeit succinctly, in the "Scientific Miscellany" column of the August 14 edition of *The Lewiston Teller*: "Dr. Carlos Finlay of Havana maintains that mosquitoes may spread yellow fever." But Finlay had yet to prove his theory.[19]

Alongside Walter Reed on his June 1900 mission were his longstanding assistant, James Carroll from the University of Maryland, and two civilian contract surgeons: Aristides Agramonte, a Cuban-American physician, and Jesse Lazear, a clinical pathologist at the new Johns Hopkins School of Medicine. After receiving his medical training in the United States, Lazear studied at the Pasteur Institute in Paris, where he developed skills as a bacteriologist. Lazear was recruited back to the United States by William Henry Welch to head the clinical pathology laboratory at Johns Hopkins Hospital. An exceedingly ambitious young scientist, Lazear had an interest in malaria and was likely aware of the work of Ronald Ross, a British army doctor working in India, and Alphonse Laveran, a French army doctor working in Algeria. Ross and Laveran had independently established that malaria was a microscopic parasite transmitted by mosquitoes. Each would receive a Nobel Prize in Physiology or Medicine—Ross in 1902 and Laveran in 1907.

Alerted to the need for a clinical microbiologist during the US occupation of Cuba, Lazear applied for the position and, upon William Welch's recommendation, was invited by Walter Reed to join the Yellow Fever Commission in Cuba as a civilian contract surgeon. Along with his wife and their infant, Lazear arrived in Cuba in February 1900, where he was joined by his medical school classmate at Columbia, Aristides Agramonte. Reed and Carroll arrived in June, and the team attempted without success to identify *Bacillus icteroides* in yellow fever victims.[20]

Although Army Surgeon General Sternberg had directed them to disregard the mosquito theory, Lazear believed that a living host was spreading yellow fever, and mosquitoes seemed the most likely culprit. When the yellow fever investigators visited Finlay in his Old Havana home, Finlay gifted Lazear mosquito eggs that he had been incubating, and Lazear used newly hatched mosquitoes from those eggs to feed on yellow fever patients in the Las Animas Hospital in Havana.

The yellow fever investigators had established their research base at the Columbia Barracks near Quemados, west of

Havana. This site was important as it was free of yellow fever. Lazear allowed Finlay's mosquitoes to bite healthy volunteers, including himself and his colleagues, but no one became sick. After only five weeks in Cuba, Walter Reed returned to Washington, DC, to attend to other responsibilities, leaving Lazear and Carroll at the Columbia Barracks and Agramonte at the Military Hospital No. 1 in Havana.[21]

The group didn't realize that an important clue had been overlooked. In 1899, Henry Rose Carter, a physician and epidemiologist with the Marine Hospital Corps (forerunner of the US Public Health Service), was assigned to serve as chief quarantine officer in Havana. Previously stationed in southern Louisiana, Carter made a crucial observation that after the arrival of a yellow fever patient in a yellow fever–free locale, it would be two weeks before secondary cases appeared. He termed the delay "extrinsic incubation," but he could not explain the phenomenon.

With Walter Reed out of the picture, Carroll and Lazear continued to test "loaded" mosquitoes—mosquitoes that had bitten yellow fever patients at the Las Animas Hospital in Havana. Beginning on August 11, two healthy volunteers were bitten, followed by six other volunteers who were tested at intervals of two to five days without effect. Expecting the same result, James Carroll allowed himself to be bitten by a loaded mosquito that had previously bitten four different yellow fever patients. Carroll became ill two days later and was hospitalized with a severe case of acute yellow fever. Carroll recovered but never fully regained his strength. He died in 1907 at age 53.

The same day Carroll was hospitalized, a young soldier, Private William Dean, volunteered to be bitten by loaded mosquitoes. Dean had remained at Columbia Barracks since arriving from Ohio, meaning that he had no yellow fever exposure. Dean was bitten by the same loaded mosquito that had bitten Carroll. Six days later, Dean developed a less severe case of yellow fever, from which he recovered.[22]

Lazear had done it. Energized by the successful demonstration that Carroll and Dean were infected with yellow fever by mosquitoes, Lazear wrote to his mother days later. "I rather think I am on the track of the real germ, but nothing must be said as yet, not even a hint. I have not mentioned it to a soul." Back in Washington, DC, Walter Reed received a cablegram informing him of Carroll's near-fatal illness. While it is believed that Reed ordered Lazear to stop human testing, no record has been found of such an order. Nevertheless, testing was thought to have ceased by August 31, the day that Carroll was hospitalized and Dean was bitten by an infected mosquito.

Ten days after he penned the letter to his mother, Jesse Lazear became severely ill with yellow fever and had to be hospitalized. Lazear claimed to have been accidentally bitten by a wild mosquito while at the Las Animas Hospital in Havana a few days earlier, on September 13. Lazear's condition rapidly deteriorated, and he took his last breath on September 25, 1900. He died at age 34, leaving behind a wife and children, a tragedy that confirmed Finlay had been right all along. A bronze historical plaque in memory of Jesse Lazear at the Johns Hopkins Hospital in Baltimore reads: "With more than the courage and devotion of the soldier, he risked and lost his life to show how a fearful pestilence is communicated and how its ravages may be prevented."

After his death, Lazear's laboratory notebook mysteriously disappeared without explanation, and with it, any documentation that Lazear may have become infected on purpose. Years later, a cryptic note in Lazear's handwriting was found on page 100 of an unused laboratory notebook, strongly suggesting that on September 13, Jesse Lazear had intentionally tested himself (referred to as "Guinea Pig No. 1") with a bite from a mosquito loaded from at least four different yellow fever patients (on August 30 and September 2, 7, and 10), which led to his death from yellow fever. None of his contemporaries believed that Lazear voluntarily infected himself, but circumstantial evidence supports the conclusion that he did. Certainly, all agreed that a mosquito caused his death.

Four weeks after Lazear's death, Walter Reed presented the early evidence for the spread of yellow fever through mosquitoes at a meeting of the American Public Health Association in Indianapolis. In a presentation titled "The Etiology of Yellow Fever: A Preliminary Note," coauthored with Carroll,

Agramonte, and Lazear, Reed detailed findings supporting their conclusion that *Bacillus icteroides* was not the organism causing yellow fever;[23] finally, they were able to refute the Sanarelli hypothesis. Reed described their experience with 11 volunteers, including Carroll and Dean, who had been bitten by mosquitoes that had fed, 2 to 13 days earlier, on sickly yellow fever patients. He also mentioned the "accidental" case of Lazear. Reed emphasized that the mosquito theory was first proposed by Carlos Finlay, who had gifted the team mosquito eggs.

The immediate response was mild, and most were unconvinced. Sanarelli's supporters voiced opposition, and *The Washington Post* dismissed Reed's presentation: "Of all the silly and nonsensical rigmarole about yellow fever . . . the silliest beyond compare is to be found in the arguments and theories engendered by the mosquito hypothesis."[24]

Determined to prove that his commission and Finlay were correct about the theory of mosquitoes as vectors, Reed returned to Cuba in November of that year and set up Camp Lazear, a quarantine area adjacent to the Columbia Barracks in a yellow fever–free zone.[25] As Carroll and Lazear had visited Las Animas Hospital and areas of Havana with endemic yellow fever, and Carroll had participated in autopsies of yellow fever victims, it was considered possible that they became infected with yellow fever by some other route. The investigators wondered how long the infectious agent needed to incubate inside mosquitoes before it became infectious to humans, so they devised a set of experiments to help answer this question—and to quell the ridicule that first Finlay and then Reed had endured from fellow scientists.

Tents were built at Camp Lazear to test a different method of yellow fever transmission. An infected clothing tent was stacked with piles of dirty towels, bedding, and clothing from yellow fever patients, its air warmed by a stove. Crusts of blood and unnamed bodily fluids festered on the fabrics. Three volunteers spent 12 nights on cots amid the dirty clothing. But as Reed's team dumped even more soiled bedding into the tent, the sight and stench of blood caused one of the study volunteers to run outside to vomit.

A second tent was split with a wire divider. One-half was connected to a mosquito breeding area with a large pool of water; the other half was not connected to the mosquito breeding area. Fifteen mosquitoes that had fed on yellow fever patients were let loose in one-half of the space, and Reed, with three Americans, protected by the wire partition, went to sleep across from the mosquito-ridden volunteers. Eight volunteers, including a man named John Moran, were asked to sleep in the mosquito breeding side of the space, where the 15 infected mosquitoes had been freed. Moran and other volunteers received two hundred dollars each to sleep in the tent for almost three weeks. Moran fell sick on Christmas Day 1900. Five more volunteers from the mosquito side contracted yellow fever. None in the infected clothing tent suffered illness, and Reed and the men on his side of the wire divider remained in good health.

In celebration of this conclusive evidence that mosquitoes spread yellow fever and that Finlay's early experiments had not allowed sufficient incubation time for the virus to become infectious to humans, Reed invited Finlay to Camp Lazear to share the results. "It was Finlay's theory, and he deserves much for having suggested it," said Reed. Gorgas agreed. "His reasoning for selecting the [mosquito] as the bearer of yellow fever is the best piece of logical reasoning that can be found in medicine anywhere."

The response to this evidence linking the *Aedes* species of mosquitoes to yellow fever was rapid and profound. Staged as a military campaign and led by another US Army doctor, William Crawford Gorgas, chief sanitary officer for Havana and a close associate of Walter Reed, Gorgas quickly translated the newly identified role of *Aedes* mosquitoes in the transmission of yellow fever as well as in dengue. Gorgas and his mosquito brigades set out to eliminate the *Aedes* mosquitoes in Havana. Relatively few homes in Havana had water delivered by pipes, with many dwellings collecting rainwater in cisterns that served as active mosquito breeding sites. Applying a thin oil layer to the surface successfully eliminated mosquito larvae. The results were dramatic. From September 1901 until June 1902, not a single case of yellow fever occurred in Havana. Gorgas went on to eliminate yellow fever and malaria from Panama, allowing for the completion of the Panama Canal. Gorgas later became president of the American Medical Association and US Army Surgeon General.

A NOBEL SNUB

Reed may have been careful to emphasize Finlay's role in the discovery, but Finlay was insufficiently recognized by the wider community for his contribution. As specified in his will, Albert Nobel, the inventor of dynamite, left his fortune to endow an international award each year for discovery in physiology or medicine. Separate Nobel prizes were designated for physics, chemistry, literature, and peace. The recipients were to be evaluated by a closed committee in Sweden, but the nominated candidates were global. The first Nobel Prizes were awarded in 1901, making the Nobel the first truly international prize. While up to three recipients could share each Nobel, no prizes were to be awarded posthumously.

Walter Reed was nominated twice, but his Nobel nominations were disqualified, as he had died from appendicitis in November 1902, making him ineligible. Jesse Lazear was never nominated for the prize. During the years 1905 to 1915, Carlos Finlay was nominated seven times without success, much to the chagrin of the Cuban people.[26] Of note, four of Finlay's nominations were submitted to the Nobel Committee by Alphonse Laveran of the Pasteur Institute, who was awarded the 1907 Nobel for discovering mosquito-borne parasites as the cause of malaria. Although Finlay was never awarded the Nobel Prize, his pivotal role in conquering yellow fever did not go completely unnoticed. At the forty-third annual meeting of the American Public Health Association in Jacksonville, Florida, in 1914, the year before Finlay's death, US Army Surgeon General William Gorgas acknowledged that "no country owes a greater debt of gratitude to Doctor Finlay than does the United States."[27]

EL COMANDANTE

———

Fidel Castro remained perplexed and incensed about Finlay's Nobel snub. "Finlay—and Cuba—were slighted!" he bellowed, jabbing his index finger in the air as he spoke.

It was late afternoon, and I had just finished giving a lecture to chemistry students at the University of Havana. Fidelito had mentioned this meeting during our lunch, but I sat in stunned stillness as I was escorted through the city in an impeccably clean white Lada. It was December 15, 2011, three years since Castro had stepped down as president, and I was on my way to meet him. With me were Alan Robock, an eminent environmental scientist from Rutgers University, and two of Robock's senior Cuban scientific collaborators. All of us had handed over our cell phones to security.

The car pulled up to the International Convention Hall, and we were escorted to a small conference room. Castro was

The Biotech Revolution During Cuba's "Special Period"

Looking at Cuba, governments might think twice about imposing economic sanctions and trade embargoes. While there is little doubt that punitive measures hurt science and impede the training and achievements of researchers, Cuba's history shows how adversity and fear can sometimes fuel scientific innovation.

After the collapse of the Soviet Union in 1991, and following decades of US embargoes, Cuba was forced into what it euphemistically called a "Special Period." This time of economic crisis threatened to undo the 35 years of gains made when the Soviet Union was Cuba's most significant trading partner and source of military and political support. In the late 1980s, Soviet subsidies made up more than a fifth of the Cuban gross national product.

Even after his most critical ally fell, Castro refused to drop the ball on science. Instead, he prioritized a universal literacy requirement and insisted that Cuba's success in science would determine its overall success. "The future of our country has to be necessarily a future of men of science," he said in a speech the year after he took power.

Cuban scientists developed the world's first hepatitis B vaccine in 1988; launched a comprehensive disease-screening program for newborns that led to a dramatic decline in infant mortality; developed low-cost vaccines and anticancer injections; churned out low-cost, high-quality generic medicines; and trained thousands of doctors and nurses to offer help at home and to impoverished nations seeking aid during disease outbreaks, including the 2014–2016 West African

Ebola epidemic. All this happened while navigating meager access to scientific equipment, a dearth of chemical reagents, and difficulties procuring items essential to the practice of science.

How did Cuban scientists do it? While government salaries, including those of scientists, are low, and the number of science PhDs has been stagnant, relatively high government funding for research has been key to Cuba's scientific success. Low-paid employees are a cheap but well-trained labor force, even as low pay fuels a brain drain and some top talent is recruited to overseas laboratories.

Out of fear that embargoes would prevent Cubans from accessing the best available medicines, Castro supported the launch of the Polo Científico del Oeste, Cuba's scientific "pole," in Havana in 1992. With over ten thousand researchers at more than 50 scientific institutions, Cuba's biotechnology hub operates in a closed cycle. Each institution constitutes its own company with its own research, development, production, and marketing plan. Among these, the Center for Genetic Engineering and Biotechnology has 1,600 employees who have commercialized around two dozen medical therapies, including the hepatitis B vaccine, treatments for macular degeneration and other eye diseases, and immunotherapies for the treatment of cancer.

The Center for Molecular Immunology in Havana developed a cancer vaccine (CIMVax) that increases survival and improves well-being in patients with non–small cell lung cancers. CIMVax induces the appearance of antibodies to human epithelial growth factor, a hormone that lung cancer cells need for growth. CIMVax and other medicines are being evaluated by clinical trials in the US in collaboration with the Roswell Park Comprehensive Cancer Center in Buffalo, New York. The

Food and Drug Administration has approved testing these medicines, which have never previously been evaluated in the United States.

Cuba's scientific gains can be measured not only in patents and discoveries but in years lived. Cubans can expect to outlive Americans by three years, despite a per capita GDP less than one-quarter that of Americans. Infant mortality rates in Cuba are on par with that of European Union nations.

As embargoes are lifted, Cuba's quiet biotech revolution will likely get louder. Forging collaborations with American scientists to expand vaccine studies and cancer clinical trials could mean greater access to treatments—not only for Cubans but for the millions of people in low- to middle-income countries whose lives depend on low-cost medicines made in Cuba.

FURTHER READING

"Country Comparisons: Infant Mortality Rate," *The World Factbook* (CIA), accessed June 16, 2025, https://www.cia.gov/the-world-factbook/field/infant-mortality-rate/country-comparison/.

Drain, Paul K., and Michelle Barry. "Fifty Years of US Embargo: Cuba's Health Outcomes and Lessons." *Science* 328, no. 5978 (2010): 572–573. https://doi.org/10.1126/science.1189680.

Luis, Luis R. "Cuba's Uncertain Revenues from Medical Exports." ASCECuba.org, July 22, 2020. https://www.ascecuba.org/post/cuba-s-uncertain-revenues-from-medical-exports.

Reardon, Sara. "Can Cuban Science Go Global?" *Nature* 537, no. 7622 (2016): 600–603. https://doi.org/10.1038/537600a.

Rhodes, Nick. "Healthcare in Cuba: A Doctor's Reality." *Borgen Magazine,* June 6, 2020. https://www.borgenmagazine.com/healthcare-in-cuba/.

sitting inside, flanked by several large men in suits and fatigue-clad military personnel bearing sidearms. He stood to greet us wearing a black Adidas sports suit with a blue plaid shirt underneath. He shook hands with each of us, and although he stooped to my height and appeared somewhat frail, he smiled brightly and pointed to our seats around the table.[28] Sitting to Castro's left was his wife; to his right was Fidelito and Castro's second son, Antonio, an orthopedic surgeon and former physician to the Cuban national baseball team. Alan and I sat across the table on either side of a Spanish translator.

Castro talked at length—over three and a half hours in total—about the global contribution made by Cuba's doctors and the advancements Cuban scientists had achieved despite stringent embargoes. It was Cuban scientists who developed the world's first meningitis B vaccine in 1988, and it was Cuban scientists who had recently discovered a vaccine to treat cancer; all the while, the medical establishments in other nations mocked the idea of anticancer immunizations, he said. Castro was particularly proud of developing universal health care for all Cubans despite the post-revolution exodus of private medical practitioners who only cared for patients who could pay for treatment.

Castro complained over and over again that the Nobel committee had failed to recognize the magnitude of Finlay's achievement. He said the Nobel snub was a national slight and that Finlay's work was foundational to every yellow fever success made since 1900—including the development of the first yellow fever vaccine in 1936 by Max Theiler, a South African working at the Rockefeller Foundation, who received the Nobel Prize in 1951.

Castro went on to share his opinions of the US presidents—nearly all were extremely negative—but he did express some sympathy for President John F. Kennedy, whose assassination Castro claimed had been coordinated by right-wing domestic terrorists. Castro reflected on Cuba's dependence on Soviet shipments of goods and supplies, but he made clear that he did not consider all Soviet demands binding. Regarding the 1962 Cuban Missile Crisis, he objected to the installation of nuclear-armed missiles in Cuba and said that he sent Che Guevara to Moscow in an unsuccessful attempt to intercede. For a second language, Castro maintained that Cubans should learn to speak English, not Russian.

At times, the conversation was casual, entertaining, even surprising. Castro described his childhood self as an incorrigible youngster who challenged his teachers and impressed other children with his extensive vocabulary of curses. He was born out of wedlock, he said, and observed the huge differences in lifestyle enjoyed by his father, a prosperous landowner, and his mother, a servant. Curious about my scientific work on aquaporin water channels, Castro asked how much water he should drink each day. He seemed aghast that I hadn't written a book and said he could easily read all 160 of the scientific papers that I had published at that point, although he asked if I could send them to him by mail, given Cuba's slow internet connection. He advised us to read *Lincoln* by Gore Vidal and, in a follow-up letter, advised me to read *In Confidence* by Anatoly Dobrynin (an alternative perspective on the Cold War, written by the Soviet ambassador to the United States from 1962 to 1986); and *In Sickness and in Power* by Lord David Owen, a physician and UK foreign minister describing modern leaders who governed despite serious illness.[29]

Alan Robock and I left the meeting amazed that we had spent the evening in the company of Fidel Castro. While Castro had discussed many subjects, we had chosen not to directly ask whether he had any regrets—and Castro didn't share with us remorse for any of his decisions or actions. As we motored back to the hotel, Fidelito was silent and spoke only once to request that we not talk to journalists until after we had returned to the United States. This seemed sensible at the time, but looking back, it seems likely that the visit had been sparked by the curiosity of Fidel Castro, who, even in his declining years, sought attention from the world beyond his native Cuba.[30]

Three years after that meeting, on the centenary of Finlay's death, I was invited to join Dean Michael Klag and the Health Advisory Board of the Johns Hopkins University Bloomberg School of Public Health to formally acknowledge Finlay's and Lazear's contributions to yellow fever research. At a Cuban

Academy of Sciences ceremony, a bronze historical plaque bearing the names of Carlos Juan Finlay and Jesse Lazear was unveiled. The unveiling was aired on state television, and days later, as I strolled through a park in Havana, people waved and said they recognized me from watching the ceremony on television. My hope was that more formal recognition of Finlay's work by an American institution and an American Nobel laureate might help allay some of the resentment that had built up over the years. Based on repeated invitations and correspondence from the Cuban Embassy in Washington, DC, signs of renewed friendship were evident, and I felt hopeful about the future of American-Cuban scientific relations.

HAVANA SYNDROME

Fidelito and I kept in contact and met whenever I was in Havana. He brought his son Fidel Antonio, who also trained as a physicist, to visit me in Baltimore and meet with faculty and students at Johns Hopkins University in March 2016. When Castro died later that year, at age 90, his death did not come as a surprise. Cuba entered a period of mourning and adjustment. While no obvious signs of increased social instability were evident, a series of unexplained illnesses or possibly attacks on US diplomatic personnel began the same month that Castro died.

The victims of these mysterious incidents complained of strange noises, dizziness, lightheadedness, and mental fatigue. It seemed particularly ironic to me that these destabilizing events should occur when US-Cuba relations were blossoming.[31]

Almost 30 years earlier, in 1987, a series of unexplained, spontaneous glass shatterings were noted by US diplomatic personnel in Havana. First, a mojito glass cracked in the hands of David Rose as he toasted his brother-in-law, Jay Taylor. As the highest-ranking American diplomat in Cuba, Taylor lived with his family in the US ambassador's magnificent residence in Havana—a limestone mansion on a five-acre plot continuously guarded by US Marines. Taylor was entertaining guests in the ambassador's residence when a vase shattered spontaneously. A few days later, toward the end of Christmas week, a drinking glass cracked in the hands of Taylor's grandson, Myles, and sliced the 3-year-old's lip open.

Jokes were made about "Marxist-manufactured glass" and poor-quality engineering. But after the vase shattered, Taylor had other suspicions. Taylor was a Marine Corps veteran who had served a tour in Vietnam. He later served as deputy assistant secretary of state and on the White House National Security Council before arriving in Havana. He had stoked the ire of some in the Reagan White House with his deep desire to advance US-Cuba relations. Still, Taylor suspected that his Havana residence was bugged and that Soviet-made sonic or directed-energy devices had been used to crack the glasses and vase.

He spoke to Jason Matthews, the CIA Havana station chief during Taylor's tenure, and learned that Matthews was experiencing a similar phenomenon. He had found a glass table cracked in his home one day. Some mornings, Matthews confided, he found cigarette butts in a living room ashtray—a sign that brazen intruders had visited.

When Taylor approached Cuban foreign ministry officials, they insisted that the supposed attacks did not come from them. But Taylor and Matthews had heard of sonic devices allegedly used by the Soviets to deploy microwaves against staffers at the US embassy in Moscow during the Cold War. The pair believed similar devices had been used against them—even if Taylor's plans to advance US-Cuba relations were stoking positive sentiments among Cuban officials. Taylor arranged for an American security team to sweep the ambassador's residence in which the glasses and vase had shattered. The security team swept the house for devices and bugs. They found none. The case was closed.

The timing of the alleged glass-shattering attack on Christmas 1987 coincided with the signing of the Intermediate-Range Nuclear Forces (INF) Treaty by Presidents Ronald Reagan and Mikhail Gorbachev in Washington, DC. This made it a pivotal time for Cuba; its strategic importance was reduced, raising uncertainty about continued support from the Soviet Union, on which Cuba was economically dependent.

Twenty-nine years later, when a medical mystery struck US diplomatic personnel in Havana, no glass shattered, but one undercover CIA officer told a US embassy nurse that he was suffering intense dizziness, crushing headaches, and feelings of pressure and strange sounds in his head while inside his Havana home. The illness followed weeks of home invasions and surveillance by Cuban operatives, not unusual given that Cuban intelligence officers, known as "entry teams," often spied on American diplomats to discover which embassy staffers were secretly working for the CIA.

Less than two weeks later, the same CIA officer fell ill again with the same symptoms. Then, a few weeks after that, two more undercover CIA officers stationed in Cuba began to feel ill, effectively debilitating the entire US CIA operation in Havana, as the bureau typically housed three CIA officers and one station chief at a time. While the chief was unaffected, the three officers said they heard loud chirps of cicadas resonating through their heads and felt waves of pressure when they worked indoors. They could not shake the disturbing sounds or feelings until they stepped outside or opened an external door or window. When the CIA sent additional staff to Havana, two fell ill. Then foreign service officers and other diplomats stationed at the US embassy in Havana reported similar symptoms. Some said they felt as if they were being hit with a targeted energy beam. Others experienced forgetfulness, dizziness, loss of balance, and intense pressure inside their heads. Previously healthy adults reported walking into doors and struggling to make sense of words on a page.

Psychiatrists consulted psychologists; military doctors deliberated with neurologists. Brain scans were analyzed, and the possibility of concussions was considered—except that the symptoms, which mirrored concussions among soldiers who suffered bomb blasts, had no apparent etiology among diplomats. Doctors could find no signs of impact. "Concussion without concussion" was how one professor of neurosurgery, Douglas Smith, at the University of Pennsylvania, described the syndrome. However, Smith's colleague, Kenneth Foster, in the School of Engineering and Applied Science, remained skeptical about directed-energy weapons as a causative agent: "You might as well say little green men from Mars were throwing darts of energy."

When American officials asked the Cubans to stop the harassment, the words of President Raúl Castro to the director general of the American Foreign Ministry echoed what Fidelito told us at lunch in Old Havana: "It's not us. . . . We need more information from your government to help solve it," the elder Castro had told him. That conversation with Fidelito took place almost a year after the first American officers suffered strange symptoms. We were attending a research conference jointly sponsored by the AAAS, Johns Hopkins Malaria Research Institute, and the Instituto de Medicina Tropical Pedro Kourí in Havana. While the conference was focused on malaria and dengue fever, the suspected attacks came up when Fidelito and I had lunch in Old Havana with the new CEO of the AAAS,

Rush Holt, a plasma physicist and former eight-term Democratic congressman for New Jersey.

Unlike his father, Fidelito always projected shyness, but he seemed unusually serious this time. He brought up the subject of the new mystery illness, an illness so transient, localized, and abstruse that it was called "The Thing" by befuddled doctors and scientists but referred to as the "Havana Syndrome" by the US news media.

"The Americans want answers from us, but we don't know what's going on," Fidelito said. He said that the Cubans wanted help from the FBI to investigate and solve the mystery. He did not speculate about who might be responsible.[32]

Cases kept occurring. By 2022, at least a thousand cases of Havana Syndrome had been reported to the US government. More than a hundred of those affected were American diplomats, but not all who fell ill were stationed in Cuba. Spies, officers, and even high-ranking CIA officials stationed in or visiting Russia, India, Switzerland, Serbia, France, Austria, Kyrgyzstan, Germany, the United Kingdom, Australia, Taiwan, China, Uzbekistan, Georgia, and Colombia described similar symptoms. Vice President Kamala Harris's 2021 trip to Vietnam was delayed when new reports of Havana Syndrome among US personnel in Hanoi were reported. Perhaps most curious of all, in 2020, four cases of Havana Syndrome among Trump White House officials were reported, including two among staffers who said they were attacked while on the Ellipse in Washington, DC.

Why were Americans under attack in Cuba and elsewhere just as President Barack Obama was restoring diplomatic ties and lifting sanctions? In 2016, the same year that Havana Syndrome emerged, President Obama flew to Cuba to declare, in Havana's Gran Teatro, that he was there to "bury the last remnant of the Cold War in the Americas." Raúl Castro had smiled in the audience. But as political relationships were seemingly on the mend, the medical mystery meant that American officers and agents were being pulled from Havana. Some were forced to leave so quickly that they were unable to pack personal belongings.

The context of President Obama's diplomatic approach might have played a role. When the president announced the restoration of diplomatic relations with Cuba, it was assumed that Hillary Clinton would succeed him and continue his work. Instead, Donald Trump won the 2016 election. Seventeen days later, Fidel Castro died.

Negotiations with Cuba had begun in secret in November 2012, soon after Obama was reelected. The president had asked Ben Rhodes to begin negotiations with the Cubans. Two years later, in December 2014, when Obama and Raúl Castro announced that the United States and Cuba were reestablishing diplomatic relations, their declaration may have sparked the ire of Russian or right-wing factions within Cuba who stood to gain if the US-Cuba conflict was prolonged. Or it might have been that Russia's president, Vladimir Putin, may have interfered since the United States and Russia were at odds over Russia's intentions toward Ukraine. In both the Cuban and US governments, the question was asked: Were these attacks the work of the Kremlin?

Havana Syndrome was initially thought to be caused by sonic attacks and, later, the work of targeted microwaves, the kinds that the Soviet Union had purportedly used during the Cold War. But there was no evidence that such weaponry existed, and the medical imaging of more than a hundred patients did not align with what would be expected in such attacks. Another theory—unpopular with the CIA—was presented: Perhaps there was no tangible causative agent. Some psychologists and neurologists proposed that Havana Syndrome was a type of mass psychogenic illness—the symptoms were real, but there was no foul play. Instead, a group of people had begun to feel ill at the same time, absent an environmental or physical etiology, a hallmark of mass psychogenic illness.

In February 2022, the Office of the Director of National Intelligence made a declassified summary public. An expert panel had convened to review the data and investigate five potential mechanisms causing Havana Syndrome, including chemical and biological agents, ionizing radiation, acoustic signals, natural and environmental factors, radio frequency, and other electromagnetic energy. The panel ruled out four causal mechanisms and found that Havana Syndrome may "plausibly" be explained by pulsed electromagnetic energy and ultrasound used in close proximity to a target.[33] "Using non-standard antennae and techniques, the signals could be propagated with low loss through the air for tens to hundreds of meters, and with some loss, through most building materials," the report said. Regarding mass psychogenic illness, the panel maintained that the latter explanation alone was insufficient, although it might have been responsible for exacerbating or prolonging the illness experienced by some diplomats.

Unlike the February report, the findings of a CIA report published a month earlier explored potential bad actors as instigators of Havana Syndrome. But the CIA report found that most incidents were unlikely to have been caused by Russia or another foreign power such as Cuba or China. Instead, it said that stress, undiagnosed medical conditions, or other unknown environmental factors likely caused the majority of the one thousand cases. The findings angered some groups of diplomats who say they continue to suffer the effects of Havana Syndrome. "The CIA's newly issued report may be labeled 'interim,' and it may leave open the door for some alternative explanation in some cases, but to scores of dedicated public servants, their families, and their colleagues, it has a ring of finality and repudiation," said a statement published by one group of Havana Syndrome sufferers.[34]

No evidence of microwaves has been found, no sonic attack or directed-energy devices detected. Still, when CIA Director William Burns visited Russia in 2021 to discuss Moscow's actions on the border of Ukraine, Havana Syndrome was discussed. Burns warned the Kremlin that if Russia were found to be involved with the attacks against US personnel, there would be diplomatic repercussions.

The medical care of returning US foreign service personnel had been delegated to several medical centers throughout the United States including Johns Hopkins Hospital. Physicians were charged with establishing therapeutic routes to recovery, but they were instructed not to undertake research studies seeking forensic evidence for the root cause of the illness. When I spoke to senior physicians at Johns Hopkins, I was told that the cases were compelling in terms of real human suffering, but the range of symptoms varied considerably, with no single feature that fit every case of the syndrome. The medical community now refers to each case of individual suffering as an "anomalous health incident." Evidence does not support Havana Syndrome being caused by a single clinical entity or energy weapon, nor does it implicate a foreign adversary.

One of my fellow physicians told me that he could not dismiss the possibility that the initial cluster of embassy staff in Havana may have experienced persistent postural perceptual dizziness (PPPD), a type of chronic dizziness that feels worse when standing. The cause of PPPD isn't known, but it's thought to be triggered when the nervous system reacts to busy, crowded, and loud environments. When I asked the doctor what might explain the spread to CIA agents at other US embassies, he said that psychogenic anxiety among agents in high-stress positions might be to blame. Curiously, the incidence of cases declined after the Russian invasion of Ukraine.

Would the mystery of Havana Syndrome take us back to the days of the embargoes? I couldn't visit my colleagues in Cuba while President Trump was in office, partly because the rules were too prohibitive. Collaborations stalled, and the fresh hope imbued by the Obama White House was dampened. The plaque I helped to unveil still hangs in the Cuban Academy of Sciences. We attempted to bridge a decades-long divide and acknowledge the critical role of Cuban scientists in understanding infectious disease, but diplomatic relations can take a turn at any moment, leaving science and scientists in the balance.

My trip to Havana in August 2017 was to be my last visit to Cuba until after the COVID-19 pandemic, and it was the last time I would see Fidelito. Although he was a respected nuclear physicist and prominent leader in Cuban science, living in his father's shadow couldn't have been easy. Fidelito always politely asked about my health, but little did I realize that he suffered from severe clinical depression. During the 14 months following the death of his father, Fidelito's depression became significantly worse, requiring hospitalization. In February 2018, Fidelito died by suicide. The loss of a proponent of scientific collaborations between Cuba and the United States was both personal and far-reaching.[35]

CHAPTER 2

The Islamic Republic of Iran
ATOMIC DIPLOMATS AND
ACADEMIC CONNECTIONS

A HUMAN CHAIN FORMED around the mountainside nuclear bunker. Hundreds of young people, mainly students from Sharif University of Technology, and a few white-haired professors, chanted, "Fordow is our heart!" They jostled against one another on a winter's day in 2013 to defend their nation's nuclear program. Among them was Iran's nuclear chief, Ali Akbar Salehi, the Iraq-born, Massachusetts-educated director of the Atomic Energy Organization of Iran (AEOI).

Fordow Fuel Enrichment Plant, the formerly secret bunker built deep inside the mountains 80 miles south of Tehran,[1] was home to an even more secretive uranium enrichment program. Inside the facility, Iranian scientists labored to enrich uranium ore to yield the uranium isotope used to generate electricity or bombs. Decades of political posturing, crippling economic sanctions, and signed-then-breached peace treaties converged around the mountains that day as the fate of Iran's nuclear program—and world peace—were set to be negotiated by politicians in Switzerland the next day. At the heart of the conflict: A silvery-white metal produced by supernovae more than 6 billion years ago, now buried hundreds of feet deep in the Earth's crust.

Inside the Fordow bunker, scientists were in the business of enriching uranium to make more U235 than could be mined from the Earth's surface.[2] Commercial power plants need reactor-grade uranium, which is enriched so that the concentration of U235 is increased four- to sevenfold—up from 0.7 percent to 3 or even 5 percent. But making a nuclear bomb requires weapons-grade uranium, which contains a mammoth 80 to 95 percent U235. Weapons-grade uranium could also be called medical-grade uranium since 90-percent-enriched U235 is used to make nuclear medicine isotopes, which are crucial in measuring blood flow and lung function and for medical research.

In its natural state, uranium is mostly a mix of two atoms, or isotopes, each with a fractionally different weight. Uranium 238, or U238, the most abundant kind of uranium, makes up 99.3 percent of uranium ore found within mines in places like Kazakhstan, Russia, Namibia, and Greenland.[3] U235 makes up only 0.7 percent of the mined rock, but it is essential for creating nuclear energy. The numbers 238 and 235 refer to each atom's mass, calculated by combining the number of protons and neutrons inside its nucleus. By definition, uranium's nucleus must contain 92 protons; one additional proton would make it neptunium, and one less would make it protactinium, entirely different elements that are not as useful in nuclear re-actors. The number of neutrons, on the other hand, distinguishes one type of uranium from its more reactive cousin.

U235, the less abundant uranium isotope, is the key to a successful nuclear reaction, the kind that gives off the energy to produce heat, power, or catastrophe. Like its more plentiful companion, U235 has 92 protons in its nucleus, but it contains three fewer neutrons—143, to be exact. This combination of 92 protons and 143 neutrons gives U235 its atomic mass of 235 and the added and highly desirable quality of being "fissionable." Compared with U238, which cannot be split and cannot spark a nuclear reaction, U235 is a powerful nuclear fuel. Inside a nuclear reactor, the nucleus of U235 undergoes nuclear fission: It splits and releases heat and extra neutrons. The extra neutrons lead to the splitting of even more atoms, triggering and sustaining a nuclear chain reaction.

Uranium enrichment technology dates back to World War II, when British, American, and Canadian scientists involved with the Manhattan Project were tasked with creating the first atomic bomb. Beginning in 1942, the scientists devised four ways to enrich uranium,[4] ultimately developing the knowledge used to detonate atomic bombs on the Japanese cities of Hiroshima and Nagasaki in 1945, killing hundreds of thousands.[5]

One method for enriching naturally found uranium involves spinning the metal in gaseous form, known as uranium hexafluoride, at high speeds with heat inside centrifuges so that the slightly lighter U235 can be separated from the redundant and heavier U238 isotope. Because U235 is present in such small amounts in nature, no single centrifuge can enrich uranium in the quantities needed for electricity or weaponry. Instead, a series of centrifuges is needed, each one separating U235 from U238.

Inside Fordow Fuel Enrichment Plant, two enrichment halls housed three thousand such centrifuges, each configured in a set of eight cascades. Ascertaining the exact level of uranium enrichment happening inside Fordow depended on whom you asked. When news of the Fordow plant was first released by Iranian officials in 2009,[6] they said the centrifuges enriched uranium to 5 percent. Experts countered that Fordow's setup was not suited to that level of reactor-grade enrichment. (Iran later said it produced highly enriched uranium, which contains up to 20 percent U235. But even 20 percent enrichment would have posed a threat since that level of U235 represents around 90 percent of the effort needed to make weapons-grade uranium.)

When Iran announced its enrichment activities at Fordow in 2009, the news came weeks after Western leaders had published a letter proclaiming that the bunker had been under surveillance for some time. Formerly a tunnel facility and a base for Iran's paramilitary force, the Islamic Revolutionary Guard Corps, the Fordow Fuel Enrichment Plant was Iran's second uranium enrichment site; the first was the Pilot Fuel Enrichment Plant in the city of Natanz in Isfahan Province, designed to house fifty thousand nuclear centrifuges.[7] Iran said the Fordow facility was constructed in late 2007, while the United States claimed that its surveillance flagged construction at the site a year earlier.

The students forming a human shield around the Fordow nuclear bunker in 2013 linked arms and chanted solidarity slogans in defense of Iran's scientific achievements.[8] They had timed their voices to be heard just days before Iranian officials would meet with leaders from the United States, the United Kingdom, France, China, Russia, and Germany (the P5 + 1) to discuss the particulars of a nuclear deal in Geneva, in which U235 enrichment at Fordow was pivotal.

Less than two weeks later, a historic deal was struck: Signed at 4:30 A.M. on Sunday, November 24, and hailed as the most significant accord between the United States and Iran since the 1979 Iranian Revolution, the agreement imposed strict constraints on Iran's nuclear program in exchange for relief from select sanctions.[9] Restrictions on trading petrol, gold, and aviation and car parts were lifted, and more than $4 billion of funds from oil sales were released to Iran from frozen accounts. In return, Iran promised to stop enriching uranium above 5 percent, to dilute or convert its stock of highly enriched uranium, and to immobilize its existing reserve of low-enrichment uranium.[10] Centrifuge installation was halted, and Iranian officials agreed to open the doors to their research facilities for possible daily inspections by the global nuclear watchdog, the International Atomic Energy Agency.[11]

But the historic accord was a temporary arrangement. Penned to last six months, the Geneva deal was meant to buy time until a more detailed and permanent agreement could be negotiated. And almost as soon as the celebrations—or, in the case of Israel, the lamentations—over the temporary but landmark deal were over, the trouble began.[12]

The negotiators were at loggerheads. For a year and a half, the chief negotiators, US Secretary of State John Kerry and Iranian Foreign Minister Mohammad Javad Zarif, went back and forth, developing the broad outlines of a framework. But they could not agree on Fordow's fate or how much of Iran's nuclear infrastructure should be dismantled for its nuclear program to be deemed a peaceful one by its allies and not one capable of quickly assembling a nuclear bomb.

Iran's reserves of enriched uranium were one point of contention. Obama administration officials said that before July 2015, Iran had enough centrifuges and sufficient enriched uranium to make up to 10 atomic bombs.[13] The so-called breakout time was another crucial point of nonagreement. This is the time needed to produce enough weapons-grade uranium to build a bomb. American advisors estimated that Iran had a breakout time of two to three months, while negotiators, who wanted systems in place to detect whether Iran was rushing to make a bomb, wanted reassurances of a breakout time of at least one year, if not longer.[14]

With infrastructural details and technical minutiae about centrifugal speeds on the table, it made sense for Kerry and Zarif to delegate discussion of the scientific particulars to those with expertise in nuclear physics. Salehi, a close colleague of Iran's supreme leader, Ayatollah Ali Khamenei, and president of the Atomic Energy Organization of Iran—who had joined students defending the Fordow nuclear bunker in 2013—was summoned to the negotiations. Salehi entertained the invitation but said he would join the negotiations on one condition: His sparring partner, he insisted, would have to be Ernest Moniz, the US Energy Secretary, a nuclear engineer by training.

The pair of atomic diplomats began heated discussions in a conference room at the Beau-Rivage Palace Hotel in Lausanne, Switzerland. This hotel happened to be the location of the Treaty of Lausanne, where leaders of the Ottoman, British, and Japanese Empires negotiated peace and the disassembly of the Ottoman Empire in 1923—a second attempt at peace after the first unratified treaty failed.

Moniz and Salehi talked about "SWUs" or separative work units, the standard measure of the effort needed to separate $U235$ from $U238$, a metric that tells the story of how much time Iran would need to make a nuclear bomb. But they also told stories about life in the nuclear physics department at the Massachusetts Institute of Technology and laughed at jokes about engineering professors. The Lausanne hotel was not the first time Moniz and Salehi had crossed paths. Their connection went back 40 years to the classrooms of MIT, where Moniz arrived in 1973 to take up the role of assistant professor of physics and Salehi began the PhD program in nuclear engineering, with a focus on fast-neutron reactors. Although they weren't friends at the time, their shared pedigree forged a bond that aided the scientific negotiations in Lausanne four decades later.

MIT's nuclear engineering program was famous for its academic rigor and for forging strong bonds among students who survived the demands of that department. The graduate students and their faculty developed a shared language born out of camaraderie and stress. When Moniz and Salehi took over the stalled negotiations after almost 18 months of back-and-forth by politicians, they spoke in a familiar vernacular, and tensions dissipated among the rest of the American and Iranian negotiators. In a gracious show of respect and admiration, Ernest Moniz and Ali Akbar Salehi each arrived at the negotiation site carrying gifts for the other's grandchildren. It was clear that each scientist had genuinely positive feelings for his counterpart.

ATOMIC DIPLOMACY

MIT played a crucial role in Iran's nuclear program, not only in the 2015 Moniz-Salehi negotiations, but in the 1970s, when graduate student Salehi and professor Moniz would have passed each other in the university's hallways. Back then, when the United States and Iran were allies, the US energetically encouraged Iran's nuclear program. In fact, Iran's first nuclear reactor was a gift from the US in 1957.[15] Ten years later, the US supplied Iran with weapons-grade enriched uranium as part of the Atoms for Peace program.[16]

Atoms for Peace was named for President Dwight D. Eisenhower's December 1953 speech; part Cold War–era propaganda aimed at quelling public fears about a bloody nuclear future, part encouragement to nudge NATO allies away from conventional weapons toward cheaper nuclear weaponry. The speech led to an international nuclear-sharing program through which the US hoped to demonstrate its dominance. Through Atoms for Peace, the US supplied nuclear reactors to Pakistan and Israel, as well as Iran, and shared nonmilitary nuclear technology with countries in Latin America and Asia.[17]

Atoms for Peace was a form of atomic diplomacy, the art of exploiting the threat of nuclear war to accomplish foreign policy and diplomatic goals. At first, the US embodied atomic diplomacy that showed off its nuclear arsenal, posturing used to intimidate other nations. But the Atoms for Peace program exemplified a shift from nuclear monopoly to nuclear superiority; the US was willing to share some of its scientific prowess to demonstrate its imperviousness and pre-eminence as a nuclear superpower.[18]

The story of atomic diplomacy began at least a decade before Eisenhower's speech with President Franklin D. Roosevelt's approval of the first atomic bomb in 1942 and his decision to keep the weapon a secret from the Soviet Union. Three years later, President Harry Truman met with British Prime Minister Winston Churchill and Soviet leader Joseph Stalin at the Potsdam Conference in Germany to negotiate the end of World War II and to forge a peace agreement. Truman didn't explicitly mention his nation's nuclear program, but he did tell Stalin that American scientists were developing a particularly catastrophic weapon.[19]

Just days after the Potsdam Conference, American atomic bombs leveled 90 percent of Hiroshima and Nagasaki, and Emperor Hirohito surrendered Imperial Japan. He cited the unleashing of "a most cruel bomb, the power of which to do damage is indeed incalculable, taking the toll of many innocent lives."[20]

Some say that Hirohito's surrender marks the beginning of atomic diplomacy. Others argue that such large-scale murder of civilians was not needed to force Hirohito's hand[21] and that Truman's intimation to Stalin about the existence of a weapon that could cause unprecedented destruction—a hint meant to deter the Soviet Union from spreading communism throughout Asia—marks the true beginning of atomic diplomacy. (That strategy may have backfired. Some regard Truman's insinuation of a nuclear weapons program as the reason the Soviet Union aggressively established a buffer zone of communist countries along its borders.)

Whenever it began, atomic diplomacy was deployed in the Korean War, the Cold War, and the Cuban Missile Crisis, where the work of nuclear physicists was used for political blustering and to indicate serious threats. And while nuclear contributions to Iran under the Atoms for Peace program continued through the 1950s and '60s, atomic diplomacy came to be regarded as ineffectual by the mid-1960s. President Richard Nixon may have considered detonating an atomic bomb over Vietnam but ultimately decided not to since the Soviet Union, by that time, possessed a nuclear arsenal as large and destructive as the US cache.[22]

In fact, the dystopian principle of mutually assured destruction—the idea that if one nation unleashed a nuclear attack, then the other would retaliate and both would be annihilated—came to define global peacekeeping strategies. Nuclear weapons, even the threat of them, fell out of favor as a diplomatic tool.

At the same time, protests against America's support of Iran's nuclear program reached a crescendo in the 1970s, as news of Shah Mohammad Reza Pahlavi's human rights record became apparent. The pro-American shah had been installed during a CIA and MI6–supported coup in 1953 that overthrew Mohammad Mosaddegh, a nationalist leader beloved for wresting Iran's oil industry away from greedy British executives.

But in the mid-1970s, with fears about mutually assured destruction and the perceived futility of atomic diplomacy, one kind of atomic diplomacy was still functioning. This version was negotiated not by politicians but by America's academic administrators and science professors. MIT began to host a special program for Iranian engineers and sent its staff to Tehran to interview 50 would-be graduate students. More than two dozen Iranian scientists arrived on the Massachusetts campus by 1975. Among them, Salehi was part of the last wave of Iranians to study nuclear physics in the United States. The program was relatively short-lived, although its impacts would ripple for decades and have global repercussions.

Soon after Salehi defended his dissertation on fast-neutron reactors in 1977 at MIT, American-Iranian relations were irreparably undone. The 1979 Iranian, or Islamic, Revolution saw the ousting of the pro-American shah and the establishment of an Islamic Republic ruled by the supreme leader, Ayatollah Ruhollah Khomeini. In November of that year, emboldened Iranian students seized the US embassy in Tehran and held more than 50 American staffers hostage for 444 days. Relations with the United States were severed, and joint nuclear projects came to an end.

ACADEMIC CONNECTIONS

Among MIT's Iranian cohort in the 1970s, Salehi was one of few graduate students to return to Iran after his studies. Many remained in the United States or settled in Canada. But Salehi went on to launch Iran's first PhD program in physics, led the Sharif University of Technology, and assumed the role of Iran's nuclear chief. He also returned to the United States to recruit Iranian graduates back to Iran.[23]

American universities have long educated heads of state, ministers of health, science diplomats, and attachés who have taken their American diplomas back home to China, North Korea, Iran, Venezuela, Russia, and other perceived adversaries or historical enemies of the United States. Johns Hopkins University has a rich network of such international connections, which some call the "science diplomacy incubator." Faculty at the Bloomberg School of Public Health, the medical school and hospitals, the Krieger School of Arts and Sciences, the Whiting School of Engineering, and the Nitze School of Advanced International Studies have learned side-by-side and forged lifelong friendships with scientists from around the world. One manifestation of the science diplomacy incubator links me to Iran's atomic czar: When I undergo a checkup at the Johns Hopkins Wilmer Eye Institute, the eye doctor peering into my pupils is Salehi's niece.

Four years before Salehi's nuclear negotiations would begin in Switzerland, he invited me to meet him in New York City in September 2011. Born the same year, we were both 62 years old at the time and shared the same thin stature and medium height. Salehi conceded that relations between the United

States and Iran were exceedingly difficult but insisted that Iran was not preparing nuclear weapons—and even if they did, he said, it would be years before they could produce a rudimentary atomic bomb. Plus, they had no delivery system, he assured me. After an hour and a half of conversation, Salehi asked if I would visit Iran to present a series of lectures. He suggested I bring my wife, Mary, and veteran science diplomat Norm Neureiter, formerly a science advisor to Secretary of State Colin Powell and senior advisor at the American Association for the Advancement of Science. I said yes, eager to visit Tehran for the first time since 1970, when, as a student backpacker, I was struck by the capital's modernity.

I was 21 when I first visited Tehran, intent on seeing the world before my medical studies began. I spent the winter and spring of 1970 and 1971 hitchhiking through Japan, roaming war-torn Southeast Asia, and enduring a hot, dusty summer crisscrossing India and Pakistan before ascending the Khyber Pass by bus to the eerily remote kingdom of Afghanistan. The towns and cities in Afghanistan were remarkable for the lack of women in public places and the near absence of modern automobiles or modern buildings. In the countryside, all the men were bearded and wore either a turban or wool Chitrali hat. Most men carried rifles and gazed disapprovingly at the long-haired, young travelers from Europe or Australia despoiling their land in search of promiscuity or hashish.

By the time I crossed into Iran, I was ready for modern amenities and the absence of acrid smoke spewing from hundreds of cooking fires. The rural districts in Iran appeared desolate, but the cities seemed to be living in a different century of human history. I arrived by rail from the holy city of Mashad into a Tehran that resembled Madrid. There were modern buildings and people wearing European-style clothing. I spotted a few clusters of niqab-clad women in Tehran's central train station. They reflected the cultural and economic gap between urban Iran and rural Iran.

The coup d'état of 1953 had ousted the elected prime minister, Mohammad Mosaddegh, through a campaign involving UK- and US-funded anti-Mosaddegh propaganda, including bribes to the Iranian press and paid protestors at anti-Mosaddegh rallies.[24] Mosaddegh was adored because he had nationalized Iran's oil industry, which for decades had been greedily run by the UK's Anglo-Iranian Oil Corporation. AIOC refused to share profits with Tehran, even after its equivalent, the Arabian-American Oil Company in Saudi Arabia agreed to split oil revenues with Riyadh in 1950.[25] When Mosaddegh shut down negotiations with the Anglo-Iranian Oil Corporation and nationalized Iran's oil industry in 1951, an exasperated United Kingdom turned to the United States for support.[26] What followed was a CIA-supported coup, Operation Ajax. Rented crowds and propaganda led to the fall of Iran's nationalist hero, who was jailed while the monarchy was restored. Shah

Mohammad Reza Pahlavi, reviled by many Iranians but a darling of the West, assumed power.

Huge consumers of revenue from the sales of oil exports, the shah and his family were stolidly anti-communist and maintained a lavish lifestyle; they were frequent subjects of numerous glamorous photo shoots at the palaces. The United States made no compromises in its support for the shah and his family. Enjoying fabulous wealth appeared particularly egregious as most Iranians remained poor. Ruthless control of the public was undertaken by the shah's notorious secret police, SAVAK.

But I could feel the tensions rising even during my 1970 backpacking visit. Nine years later, nationalism, fomented by the 1953 coup, erupted in a revolution that cast out the shah and replaced his government with the Islamic Republic led by Ayatollah Khomeini and supported by the paramilitary Islamic Revolutionary Guard. When the Khomeini died in 1989, he was replaced by another equally acerbic supreme leader, Ayatollah Ali Khamenei.

The US National Academies of Sciences, Engineering, and Medicine had a long-standing science engagement program with Iran meant to foster good relations and effective scientific collaborations. The program was suspended in 2009, however, when President Mahmoud Ahmadinejad was elected for a second term.

A few months after our meeting with Salehi in New York City in 2011, Norm Neureiter and I met with Hassan Vafai in Washington, DC. A professor at the Sharif University of Technology, where he worked closely with Salehi, Vafai was my age and had studied and taught civil and mechanical engineering at several American universities. Vafai had previously taken Thomas Schelling, the 2005 Nobel laureate in economic sciences, and other prominent US scientists to Iran and was eager to organize our trip, telling us excitedly that he anticipated an enthusiastic response by the deans of Iran's universities to a visit by another US Nobel laureate. But there was a flip side to his enthusiasm. US sanctions on Iran had ruined the Iranian economy, he said; the currency had recently plummeted to half its value, and Vafai himself had seen his life savings melt away. Scientific diplomacy promised a route to peace and good relations.

The visit would not be an easy one. Vafai warned us that Iran was a theocracy, with the Supreme Ayatollah having virtually unlimited power. The president of Iran had little power, and university professors had essentially none. I was worried that our visit, even with its peaceful aim of fostering diplomacy between American and Iranian scientists, could be jeopardized by the Islamic Revolutionary Guard or the agents behind a series of assassinations of nuclear engineers on the streets of Tehran. It turned out that we would be confronted by tensions over those assassinations during our visit, although it would be my wife, a preschool teacher and not one of the scientific delegation, who would have to publicly face that anger.

It didn't help my anxiety in the planning phase when I spoke with Glenn Schweitzer, who worked at the National Academies and had organized multiple workshops for Iranian scientists in the United States and in Iran. During one of his visits, scientists were roughly detained by the Islamic Revolutionary Guard. Schweitzer urged us to be extremely cautious, and Vafai decided to restrict our first trip to Tehran and to arrange for plainclothes police carrying concealed weapons to accompany us at all times, even inside our hotel.

Our official hosts in Tehran in June 2012 included Nasrin Soltankhah, a mathematician and Iran's vice president for science and technology, and Saeed Sohrabpour, director of the National Elite Foundation, Iran's equivalent of the US National Academy of Sciences. Sohrabpour earned his PhD in mechanical engineering at the University of California, Berkeley.

Our mini-delegation of three included Neureiter, my wife Mary, and myself. We were tasked with improving relations with Iran's top scientists and government leaders. Neureiter and I were asked to present scientific lectures, while Mary was asked to deliver a speech to the Diplomatic Ladies Group. We always traveled as a convoy preceded by a tall, helmeted policeman riding a huge BMW motorcycle. Sohrabpour explained that security personnel would be staying in the hotel rooms adjacent to ours and that we must be accompanied whenever we left our room. Vafai explained that strict precautions were taken because of the recent assassinations of Iranian nuclear engineers by unknown assailants, although their deaths were widely assumed to have been organized by Israel's intelligence agency, Mossad.

Despite taking precautions to avoid the hazard of being used to endorse political factions, I was surprised to see that I had been exploited in exactly that way. After my lecture to the Elite Foundation, I spoke to an Iranian reporter who quoted me in the Islamic Republic News Agency's English-language newspaper as critical of American and United Nations sanctions on Iran. The story said that I believed the majority of American scientists did not object to Iran's uranium enrichment program at Natanz, where Iranian nuclear engineers had achieved up to 20 percent enrichment. To its credit, when I clarified that my comment only referred to enrichment to the level needed for peaceful civilian purposes, such as generation of electricity or generation of isotopes for nuclear medicine, the news agency printed a correction.

Mary was asked to give a speech to the Diplomatic Ladies Group at a special meeting organized by Salehi's wife, Zahra Rad. Attending the speech were the spouses of several Iranian government ministers and wives of ambassadors from several Middle Eastern nations. Their questions reflected a genuine interest in the family life of a Nobel laureate. Mary assured them that we lived in a normal neighborhood; further, our four children had gone to public schools, and after graduating from university, all four took jobs involved with social welfare or protecting the environment.

Just as the meeting was ending, a young woman with a small child spoke loudly in English: "How would you feel if your husband was brutally murdered in front of you and your daughter?" She accused the United States of supporting terrorism. A year earlier, the young woman had watched as her husband, the 35-year-old electrical engineer Darioush Rezai-Nejad, was gunned down in front of her and their 4-year-old daughter. The room fell silent as the tearful woman sobbed inconsolably. Mary honored the silence and then said: "The terrorists want to frighten us all and keep us locked in our homes. But they have failed because we are here, and we have come to Iran to support you." Mary reassured the young widow that we sympathized with her grief and vehemently opposed terrorism in any form. Vafai, who attended the luncheon meeting, appeared relieved by Mary's comments and, later, in private, told her that her words had been helpful.

Nejad was one of five Iranian nuclear scientists assassinated between 2007 and 2021. The previous year, Majid Shahriari, a 43-year-old physicist at AEOI, was killed when assassins on motorbikes detonated a C-4 explosive on his car door. In January 2010, Masoud Alimohammadi, a 50-year-old quantum field theorist, was killed when a motorbike loaded with explosives was detonated outside his home as he left for work. Alimohammadi had been Iran's first student in the nation's first PhD program in physics—launched by Salehi himself. The young Iranian man who was later executed for Alimohammadi's death confessed to working for Mossad.

Four months after Rezai-Nejad was shot dead in Tehran, Israeli forces killed 16 Iranian soldiers and the founder of

Iran's ballistic missile program, Tehrani Moghaddam. Two months later, Mostafa Ahmadi Roshan, a 32-year-old nuclear engineer, was assassinated with a limpet bomb attached to his car door by a motorcyclist who escaped in traffic. In 2020, Iran's top nuclear scientist, Mohsen Fakhrizadeh, was killed in Iran by an AI robot operated by agents in Israel. The sixth target, Fereydoon Abbasi, head of the AEOI from 2011 to 2013, survived a 2010 assassination attempt but died in a 2025 attack.

Iranian leaders blamed Israel's Mossad, which they said had received American approval for the attacks.[27] Spokespeople for the US Secretary of State vigorously denied US involvement, but the Israeli government declined to comment. While the six victims were young and not all were the most important figureheads in Iran's nuclear program, the message delivered was a clear one: Outside forces could strike within Iran.

But Salehi countered that the assassinations had a fortifying effect on Iran's scientific community. "We have a very peculiar characteristic of our nation. Being Muslims, we are ready for any kind of destiny because we do not look upon it like you have lost your life," he told *Science Magazine* in 2015. "For our people, it's easy to absorb such things. I mean, this did not really turn into an impediment to our nuclear activities. In fact, it gave an impetus to the field, in the sense that after [the assassinations], many students studying in other fields changed to nuclear science."[28]

Our final event in Tehran was an unexpected one. President Mahmoud Ahmadinejad learned that a Nobel laureate was visiting and asked that we join him for tea on the last morning of our trip. This put us in an awkward position. Ahmadinejad had consistently and publicly taunted the United States and Israel with scathing criticisms. But Neureiter felt strongly that we had to accept the invitation even though it could be exploited and framed as evidence of our approval of Ahmadinejad. I wasn't sure what to expect of the man who said that America had entrusted itself to the devil. But in person, the president was a docile man. We drank hot tea in an ornate sitting room in the presidential residence as Ahmadinejad spoke to us through his interpreter. It was his attempt to put a positive spin on our trip, to let us know that, while he might think our nation devilish, he could be a cordial and reasonable host. He talked about nations working together as if he had an entirely different speechwriter to his 2007 visit to New York City, where, ahead of an address to the United Nations, he had called for an overthrow of Israel and the United States. He did, however, round off his address to staff and students at Columbia University that year with these remarks: "We wish for the day when pious and pure scholars and scientists run the world."

Although the trip achieved the initial objective of restoring some of the friendship and enthusiasm previously linking scientists from the United States and the Islamic Republic of Iran, continued contact and increased efforts will likely be necessary to end the sanctions against Iran and restore economic vitality. For this, both nations will need to continue with their dialogue.

The plan was to return to Iran within a year, but those hopes fell apart when visitors to Tehran were held hostage by the Islamic Revolutionary Guard.

SANCTIONS AND SCIENCE

Three years after my trip to Tehran, Moniz and Salehi ironed out the details of a nuclear agreement that had stymied Ahmadinejad and his counterparts. After days of closed-door negotiations about nuclear reactors, heavy water, enriched uranium, specialized centrifuges, and nuclear bunkers, Moniz and Salehi came to an agreement. To be freed of crippling sanctions, Iran would allow international inspectors into its nuclear facilities and stop the types of uranium enrichment needed for making weapons.[29] Specifically, Iran would reduce its enriched uranium stockpile by 98 percent so that, until 2031, it would not exceed its reserves of 660 pounds. It also promised not to surpass an enrichment level of just under 4 percent.[30]

Then there was the question of nuclear reactors. One particular plant in western Iran sparked the concern of Western leaders who worried it had the hallmarks of a weapon-making facility. Salehi and Moniz decided that Iran would redesign the reactor in Arak so that it could still function but not as a means for manufacturing materials used for military purposes.

Fordow, however, provoked strong emotions. Salehi wanted the facility to remain open; US officials wanted it shuttered. Moniz and Salehi agreed that Fordow would stay open but not as a site of uranium enrichment, at least not until 2031. The hundreds of remaining centrifuges inside the plant could be used only to produce the kinds of uranium needed for medical or scientific purposes.[31] Research and development could still take place at the Natanz facility with its thousands of centrifuges, but Fordow would be converted into a technology and research center.

In return, Iran would receive more than $100 billion in assets from overseas accounts that had been frozen for years.[32] Sanctions, which had cost the country hundreds of billions of dollars, including a loss of more than $160 billion in lost oil revenue from 2012 to 2016, would be lifted. With these conditions finely tuned by Salehi and Moniz, the historic deal, known as the Joint Comprehensive Plan of Action (JCPOA), was finally signed in July 2015. That fall, the US Congress and the Iranian parliament formally adopted and endorsed it. The American engineer and the Iranian physicist were credited with the success of the Iran Nuclear Deal. At the heart of their scientific diplomacy lay their shared American education. Had Moniz and Salehi not shared that common bond, the historic accord might never have been reached.

But the work of the scientists was unraveled by a politician. In 2018, a year after he took office, President Donald Trump withdrew from the JCPOA. More than two years of complicated negotiations came undone, and old sanctions were reimposed. While Israeli Prime Minister Benjamin Netanyahu celebrated Trump's "courageous leadership," former President Barack Obama and others warned of a growing threat of nuclear war in the Middle East. Trump said the approach to the nuclear arms race that he had taken with North Korea, one of posturing and pressure, should be applied to Iran so that a new, more restrictive deal could be signed.[33]

Iran responded to that 2018 announcement by accusing Trump of not sticking to international treaties and saying it would abide by the 2015 deal. But it had, in reality, breached the 2015 accord four times.[34] Soon after the United States killed a senior Iranian military commander in a Baghdad airstrike in 2020, a spokesman for the AEOI declared a fifth breach.[35] "Iran's nuclear program no longer faces any operating restrictions," he said. The AEOI said it would stop its series of escalating breaches only if the P5 + 1 countries lifted the new (or reimposed) sanctions that were enacted by the Trump administration in 2019.[36] "If the sanctions are lifted and Iran benefits from its interests, the Islamic Republic of Iran is ready to return to its obligations," read an AEOI statement. Plans for a new nuclear power plant were announced.

The political-scientific scuffle continued with another American foe stepping into the picture. Russia, following its invasion of Ukraine in 2022, joined Iran as a nation whose energy

Doomed to Cooperate

During a student camping trip to the Soviet Union in the summer of 1966, I saw immense plots of wheat, corn, and sunflowers stretching to the horizon. The roads to Kharkiv, the second-largest city and a major center of Ukrainian culture in the Russian Empire, were lined with blooming flowers. The city was filled with modern architecture, replacing the wreckage caused by bombs during World War II.

Russia has long envied Ukraine's large tracts of fertile farmland and the region's rich deposits of iron ore and other metals. The incorporation of Ukraine into Tsarist Russia was pivotal to the emergence of Imperial Russia three hundred years ago. The Ukrainian cities were "Russified," while the Ukrainian farmers became serfs to the Russian-speaking landed gentry. To quash Ukrainian nationalism during the collectivization of farming in the 1930s, Stalin ruthlessly starved Ukrainian farmers and replaced them with Russians. With the dissolution of the Soviet Union, Ukraine took giant steps to join the rest of Europe as a new democracy. But the rise of Vladimir Putin in Russia was spurred by the ambition to restore Russia as a world superpower. The Russian invasion of Ukraine on February 24, 2022, was a vicious, unprovoked attack intended to return Ukraine to Russian dominance.

The world erupted in horror. Joining the outrage in response to the invasion, scientists came together to protest. An open letter signed by more than two hundred Nobel laureates condemned the attacks, comparing them to the Nazi German attack on Poland in 1939 and calling for world support for Ukraine.

Another group of scientists watched the war from 260 miles above the Earth's surface. Seven astronauts were aboard the International

Space Station as Russian troops invaded Ukraine. Among them were two Russian cosmonauts, four NASA astronauts, and one European Space Agency astronaut.

Space exploration has long relied on international collaboration— even during times of war. America's civilian space cooperation with Russia began in the midst of the Cold War. In 1975, American astronauts on an Apollo spacecraft connected with Russian cosmonauts on a Soyuz spacecraft and lived together in space for nine days. But war can stall science, even science taking place hundreds of miles above a war zone. "Whenever you have trouble on the face of the Earth, it makes it more difficult, but we've weathered this before," said NASA administrator Bill Nelson. Open space is such a hostile environment that scientists are doomed to cooperate, according to one former head of the Russian Space Exploration Institute.

The International Space Station is an east–west bridge of sorts. The US side of the station supplies power, while the Russian side supplies propulsion. As the International Space Station orbits Earth, it produces atmospheric drag that pulls it down, so every now and again, the Russian side fires up its thrusters to keep the space station on track.

Kicking Russia off the space station would not only doom its path but also end much of its space research. A Russian module, known as the Multipurpose Laboratory Module-Upgrade, houses the station's experiments.

Supply chains integral to science are also deeply interconnected. The US shoots spy telescopes into space with a rocket that uses Russian engines at its base, while some European agencies use rockets that need a Ukrainian upper stage.

Human space entry requires collaboration, too, since every American astronaut must learn Russian to speak to mission control in Moscow. But the 2022 invasion of Ukraine threatened to undo the decades-long alliance, highlighted most vividly perhaps with former NASA astronaut Scott Kelly and Dmitry Rogozin, director of Roscosmos, Russia's space corporation, engaging in a Twitter argument about the war.

Outraged by news of sanctions against Russia, Rogozin shared a video produced by his agency that showed two Russian cosmonauts inside the space station waving goodbye to an American astronaut who was supposed to return to Earth with them on a Russian spacecraft. Kelly felt compelled to respond. "Get off, you moron! Otherwise, the death of the International Space Station will be on your conscience," Rogozin said in a now-deleted tweet before appearing to threaten to drop "a 500-ton structure to India and China. . . . If you block cooperation with us, who will save the ISS from uncontrolled deorbiting and falling into the United States?" Rogozin tweeted. Kelly later backed away from the Twitter beef following an email from NASA warning all astronauts that such exchanges were "damaging" to the joint space mission.

No matter what happens on Earth, scientists will have to keep things peaceful in space.

FURTHER READING

NASA. "International Cooperation." https://www.nasa.gov/international-space-station/space-station-international-cooperation/.

"An Open Letter in Support of Ukraine." *The Economist,* March 3, 2022, https://www.economist.com/letters/2022/03/03/an-open-letter-in-support-of-ukraine.

industry is sanctioned by the United States.[37] Restrictions and common enemies can also serve as fuel for forging strong collaborations. Since the sanctions, the two nations banded together to increase trade by more than 80 percent. In 2021, Iran and Russia surpassed a record-breaking $3 billion in trade, vowing to more than triple that record in coming years. Whether science can heal these deepening fissures remains to be seen. But the hope remains that academic connections and shared scientific interests can nurture a new wave of atomic diplomacy and global peace.

CHAPTER 3

The Democratic People's Republic of Korea

EXISTENTIAL THREATS FORGE COLLABORATION

"THAT DAMNED LINE"

Two Americans carved a single Korean nation into two at the end of World War II. During an all-night meeting on August 10, 1945, four days after the United States dropped a bomb on Japan, the Allied forces decided what to do with Japan's colonies. American Colonels Charles Bonesteel and Dean Rusk spent the night staring at a map of Korea, wondering where to draw a dividing line that would satisfy Soviet and American appetites.

They were looking at the wrong map. The chart they wanted was colored with the contours of Korea's provincial boundaries, but that scroll couldn't be located.[1] Instead, Bonesteel and Rusk drew a line along a circle of latitude 38 degrees north of the equator. This 38th parallel split Korea into the Republic of Korea in the south and the Democratic People's Republic of Korea (DPRK) in the north. It was supposed to be a temporary division, a 250-kilometer line stretching east to west that would remain in effect for just enough time to allow Soviet and American troops to oversee the ousting of Japanese forces.

Five years later, the line remained. North Korea breached the 38th parallel, igniting a war between the newly carved-out nations. After two years of war, that arbitrary line became known as the military demarcation line, or the demilitarized zone (DMZ), with a one and a quarter–mile safe zone on either side, part of the "truce" that ended the Korean War. Except that war never ended, and the DMZ is anything but demilitarized. It remains one of the most militarized zones in the world, riddled with landmines and traps, with 750,000 heavily armed DPRK forces on one side and 450,000 South Korean and 20,000 American troops on the other.[2]

South Korea's first president, Syngman Rhee, believed that the two nations should be reunified as quickly as possible. One of his staffers referred to the 38th parallel as "that damned line."[3] But, more than seven decades later, political tensions between North Korea and its rivals have worsened. To live on the northern side of the 38th parallel is to live a shorter, sicker, and hungrier life.[4] Compared with their South Korean neighbors, North Koreans die a decade earlier, are three to six inches shorter, and are around six times more likely to suffer tuberculosis. The Red Cross estimates one in two babies born in North Korea dies before the age of 5 years. During the four-year famine that began in 1994, one in ten North Koreans—upwards of 2.5 million people—starved to death.[5]

MAKING CONTACT WITH NORTH KOREA THROUGH SCIENCE

North Korea has remained isolated, with few diplomatic relations and essentially no respectful contact with the United States since its inception in 1945, but its leadership has shown a strong interest in science and technology. With this in mind, faculty at Syracuse University in New York began to develop an academic partnership with North Korea's leading technology university, Kim Chaek University of Technology, in 2001.[6] It was the first significant partnership between American and North Korean academic institutions.

This unique collaboration focused on developing a digital science library to give North Korean scientists access to open-source software and scientific literature that had been available everywhere in the world—except North Korea. A second area of interest was the English-language training of North Korean scientists. Over the next decade, the partnership facilitated more than a dozen events in Pyongyang, Syracuse, and Beijing, with students and faculty from both institutions working together and publishing scientific reports as colleagues. The collaboration focused on branches of information technology including machine translation and decision support systems, as well as the design and maintenance of laboratories.

While scientists began to seed the work of cross-border collaborations, science journalists, along with most other Western journalists, remained forbidden from entering North Korea. But in 2004, one American journalist stationed in Beijing for *Science* (a publication of the American Association for the Advancement of Science [AAAS]) was invited to visit North Korea as a guest of the Ministry of Culture.

With his strong background in genetics and biophysics, reporter Richard Stone's landmark trip allowed an American science journalist to visit several life science research sites and sample North Korean culture. Stone was impressed with how science was seen as a means of enhancing the well-being of North Korean citizens, but he observed an equal emphasis on ideology and a strong military. He also witnessed evidence of the country's recent history of famine and ongoing problems with malnutrition and stunting.

Two years after Stone's visit, the Syracuse University–related project was officially launched, and in January 2006, the Syracuse–Kim Chaek digital library was opened in Pyongyang.[7] The launch sparked conversations about bigger and more frequent exchanges between the two campuses and anticipation of new scientific collaborations between North Korea and the United States. But not long after, politics got in the way. In the fall of 2006, the Six-Party Talks—a series of multilateral negotiations between North Korea, South Korea, the United States, China, Japan, and Russia focused on dismantling North Korea's nuclear program—were stalled when suspicions that North

Korea was cheating on its nuclear promises were confirmed.[8] In October, just 10 months after the Syracuse–Kim Chaek digital library was cheerfully opened, an underground explosion in northeastern North Korea was detected and announced by the US government. The blast released energy equivalent to one kiloton of TNT—one-fifteenth of the force delivered by the atomic bomb dropped on Hiroshima.[9] It was confirmation that North Korea was developing nuclear weapons.

The view broadcast by North Korean officials remained one of supreme confidence and *juche*, the North Korean term for self-determination and self-reliance.[10] Except, as Stone and Syracuse University scientists noted, there was sparse evidence of self-reliance across the nation. North Korea's meager economy is based on the export of coal, the supply of mercenary laborers, cybercrime, and the manufacture of contraband military equipment. On televised appearances, government officials claimed that possessing nuclear weaponry would protect the nation from military aggression by its neighbors or the United States.

* * *

Just when a friendlier future and international collaborations seemed a distant hope, music calmed the storm. In February 2008, the New York Philharmonic performed live in Pyongyang, softening tensions and lifting the spirits of North Koreans. It seemed that engagement in the arts and culture—

which contributed to those intangible elements that American political scientist Joseph Nye coined as soft power—had the power to defuse cross-border conflicts.[11] Could science engagement exert its own power?

THE SCIENCE OF BUILDING BRIDGES

Against this 60-year backdrop of political failure and diplomatic turmoil, I was invited to join a small delegation of American scientists on a rare official visit to North Korea. The US-DPRK Science Engagement Consortium planned to visit North Korea in December 2009, one month after my first trip to Cuba. As incoming president of the AAAS, I was excited to join the trip, one of the first for the organization's recently launched Center for Science Diplomacy. We had four main goals: to establish contact with North Korean scientists, befriend the students who would become future scientific leaders, find common ground with our North Korean peers, and align US and North Korean scientific interests. We believe that we have fulfilled the first two goals and made progress toward the third, yet we remain in the dark when it comes to the fourth.

We all shared a deep interest in building connections with the isolated nation and the use of science as a tool of diplomacy. Each of us was taking steps to forge collaborations with North Korean scientists on topics as diverse as water conservation and diversity in STEM. The consortium worked for months to obtain an invitation from the DPRK State Academy of Science while logistical and scientific preparations went ahead in anticipation of a fruitful visit. Despite those efforts, we were denied entry. But adding a Nobel laureate in chemistry as a senior member of the DPRK Consortium apparently increased interest. We received an invitation and visa approvals for a one-week visit to North Korea in December 2009. This was one of my earliest experiences of the Nobel Prize acting as a kind of pixie dust; the Nobel could open doors and make possible diplomatic missions that were previously denied.

* * *

Even with access granted, nothing about this visit was easy. Thick dockets of international sanctions exchanged at the highest levels of government felt tangible and present during our visit, as if the sanctions were visible and visceral on the streets and in our meeting rooms. The United States began to impose economic sanctions on North Korea beginning in the 1950s.[12] While many of those restrictions were loosened in the 1990s, the UN Security Council passed close to a dozen more sanctions beginning in 2006. These newer sanctions froze the assets of people involved in the country's nuclear program and banned the trade of military equipment. While it might seem as if sanctions are focused on stopping the export of luxury goods, such as superyachts and jewelry, they tar with a wide brush: scientific collaborations are collateral damage.

A HOPEFUL CONSORTIUM

The initial US-DPRK Science Engagement Consortium included Vaughan Turekian from AAAS; Stuart Thorson from Syracuse University; Fred Carriere from the US-based Korea Society; and Cathleen Campbell and Linda Staheli from CRDF Global (Civilian Research Development Foundation), a nongovernmental organization that works to locate and decommission Soviet-era nuclear weapons. Maxmillian Angerholzer III from the Lounsbery Foundation supported our travel. We planned for a December 2009 visit, and a request was made to the DPRK Mission to the United Nations in New York City, which approved and granted visas. But again, politics got in the way. Seven months before our hopeful departure, North Korea detonated another nuclear device, this one equivalent to five kilotons of TNT.[13] Next came missile launches.[14] Our work was becoming increasingly diplomatic, as we were asked to brief the staff of Senator Richard Lugar, chairman of the US Senate Foreign Relations Committee; the US State Department's Korea desk; and Ambassador Stephen Bosworth, Special Envoy to the DPRK.

These diplomatic meetings were encouraging, but pessimism lingered: Could our visas be canceled and months of planning go to waste? We imagined all of the scenarios that could disrupt our carefully coordinated trip: another missile launch, a 10-kiloton blast, or perhaps a crude comment by a government official triggering a new round of sanctions. Despite the DPRK's escalating show of military force, our consortium planned to stay on schedule, and we felt even greater urgency to use science as a tool of engagement. We considered our explicit goal of building bridges and nurturing diplomacy where politicians had failed. But we knew it might be impossible to speak to individual North Korean scientists one-on-one, away from the group. I expected that I might be followed constantly by men in suits, my movements tracked, and conversations with North Korean scientists monitored. Still, we proceeded with our travel plans and booked flights into Beijing, the only entry point to the DPRK.

Vaughan Turekian was clear that this was a first visit during which few tangible outcomes might be achieved. But in some ways, it was the intangibles that interested us. "There is such limited connection between the two countries that the simple act of meeting each other was a major accomplishment," Turekian said.

We landed in Beijing on December 8, 2009, ahead of a meeting with officials at the Chinese Foreign Ministry. The Chinese diplomats wished us well on our trip and emphasized the important possibilities of our mission. The Nobel Prize would be deeply respected, said Director General Zheng Zeguang, and one way to know whether our trip was a success was to see if we were invited back.

But our fears of impending hostility were realized soon after. At the Beijing airport, we were aggressively questioned by Japanese journalists who insisted on knowing whether we were traveling to Pyongyang. We declined to comment, nervous about raising expectations—or suspicions. We avoided the questions and hurriedly boarded the jet of the world's lowest-ranked airline, Air Koryo.[15] The 1960s-era Soviet Ilyushin airplane looked like it had been repainted by hand with enamel and a paintbrush.

As we descended into North Korea, I looked onto treeless mountains and vast expanses of dry, eroded riverbeds. We landed in Pyongyang next to aging helicopters and a few military planes parked behind fortified barriers. Twenty men in long, dark military overcoats and fur hats stood beside the jet's staircase and watched as we disembarked. After inspecting our passports and documents, they allowed us to meet the four hosts who would serve as our minders and trail us everywhere over the next week.

Hong Ryun-gi, a 62-year-old man with a friendly smile, greeted us and shook my hand. He was the director of international organizations at the State Academy of Science (SAOS) and had trained as a computer scientist at Kim Chaek University. Standing beside him were Pak Yong-il, a former military officer and IT scientist; Chol Kwang-hun, a senior officer; and a staff member. The men were polite and formal.

Out of respect for my designation as group leader, Dr. Hong and I were driven separately in an SUV while the rest of our group followed in a van. The roads were pitch black. Still, we passed dozens of people walking or bicycling along the roadside. At one point, two dozen or so soldiers darted across the road, pulling a large military wagon. We drove past a few billboards showing heroic DPRK soldiers and workers, and as we entered the city, a large image of the nation's Great Leader, Kim Il-Sung, smiled down at our motorcade.

We were assigned non-adjacent rooms in the 44-story Koryo Hotel in central Pyongyang, which sparkled with ornate fixtures and marble floors. We met with Dr. Hong to discuss our rigid schedule, which faced three new challenges: Universities were closed because of the global spread of H1N1 influenza, white-collar workers were expected to perform manual labor on Fridays, and Saturdays were a day of professional study. Dr. Hong admonished us never to leave the hotel by ourselves. Accompaniment by him or one of the hosts was necessary at all times. This set the tone for the remainder of the trip.

At 5 A.M. the next day, I awoke to the eerie, high-pitched tone of an echoing melody; it reminded me of the soundtrack from a 1930s horror movie. I opened the window and, through the frigid air, listened to a ghostly voice singing in Korean. "Where are you, Dear General?" is a call to all workers to labor hard for the good of their country. In the dimly lit streets far below, I could see workers hurrying in the predawn darkness.

After breakfast, we set out for the first scientific visits, passing through the tidy streets of Pyongyang, which were almost devoid of autos. As before, Dr. Hong and I were in an SUV while the rest of our group followed in a van. As we made our way to the State Academy of Sciences, passing the 170-meter-tall Juche Tower with its sculpted flame at the top, Dr. Hong explained that the tower symbolized the DPRK's philosophy of *juche*. Another slogan I learned was songun, "The army comes first," which served to remind North Koreans that when it came to the country's stretched food supply, the military would always be fed.[16]

Inside the biological branch of the SAOS, we met with cell and tissue engineers seeking to make drought-resistant plants and bacteriologists developing medical treatments. We walked past a cabinet filled with AK-47s. We stared at each other but were afraid to ask why a biology lab needed a stash of rifles. Later that day, Dr. Hong brought up the topic of guns. When he had told his grandson that he would be staying in a hotel and serving as a guide to visiting American scientists, the young boy became alarmed and urged his grandfather to "Bring the rifle!" The child had heard scary things about Americans.

At the Institute of Hydraulic Engineering, we hoped to learn about water purity, arsenic contamination, and waterborne diarrheal diseases, but the scientists couldn't answer our questions. They said that their entire focus was on flood control. A few of our meetings went this way: North Korean scientists were eager to speak with us but unable to answer our questions or speak broadly.

Our brief visits to the nation's top academic institutions revealed small truths about life in North Korea. At the Institute of Thermal Engineering, an 80-year-old professor gave a presentation that was powered by a small generator because the institute did not have electrical power or central heating. North Korea has no significant oil or natural gas, and the need to conserve imported fuel was readily apparent. I wondered whether the Red Cross Hospital also lacked heating. The small complex of medical buildings looked over a central courtyard and was very tidy but conspicuously lacking in activity. I saw several workers wearing hospital scrubs. They had spades in their hands and were busy with gardening work.

We rode back to Pyongyang from the University of the Sciences in Unjong Valley, a 30-minute drive north of the capital. Armed guards stationed at the entrance to the tunnel, which connected the Korean countryside to the capital, policed the entry of the rural famished poor into Pyongyang, where food rations were stockpiled. An estimated 40 percent of North Koreans—around 10 million people—suffered severe food insecurity in 2019, according to the United Nations World Food Program.[17] Official rations provided 300 grams (10.58 ounces) of food a day to each person, which caused the starving to sometimes flee the countryside in desperate search of food—if they could get past the guards.[18]

As we approached the tunnel, guards in long overcoats and fur hats stopped our vehicles and whispered into their walkie-talkies. Dr. Hong appeared annoyed and stepped out of the SUV to speak with the guards. I was left alone with his subordinate,

Pak Yong-il, a cheerful younger man who seemed eager to chat. Mr. Pak spoke excellent English, but despite his articulate diction, I was surprised when I mentioned the search engine Google, and he replied, "What is Google?" When I attempted to explain my background, he smiled and said, "Peter, I know all about you." This immediately got my attention.

I told Mr. Pak I found it surprising to see so many soldiers throughout the city and countryside. "We have the third-largest army in the world," he said, gesturing toward the guards circling our car. With a population of more than 25 million people, the DPRK had an army of at least one million active-duty soldiers and another four million reservists.[19] Unless deferred by specific academic pathways, most adult males in North Korea are conscripted to serve 10-year enlistments to generate this military force.

I watched as the guards glared into our SUV. Mr. Pak continued: "But when we spend all of our resources guarding our country, who is going to do the work?" Too wary to comment on the DPRK military, I sat in silence as Mr. Pak began talking about nuclear weapons—the first and last time the topic would be mentioned during our week-long visit. "We need an atomic weapon," he said. "We need an atomic weapon because it would be a far more efficient deterrent against our enemies compared with an army." When I asked about the recent announcements that the DPRK had successfully developed and tested missile launches not long before we began our trip,

Mr. Pak acknowledged that he and his colleagues were also very surprised by the event as there had been no prior announcements.

* * *

Mr. Pak's soliloquy felt staged at first—it was almost too precisely stated to be unrehearsed. But as he spoke, I felt for the first time that week that he saw me as a reliable and objective visitor, a fellow scientist, albeit a foreign one, who was curious about his country and who came with no intention of destroying it. I watched as the tunnel guards continued to interrogate Dr. Hong outside the car while Mr. Pak justified his country's need for an atomic weapon. It began to sound like he was offering a sincere explanation, but then I wondered whether his long pauses were designed to provoke a response. It seemed likely that he was an intelligence officer, and I didn't want to say anything that could be interpreted as hostile or even critical of the DPRK. We sat in silence for several minutes until Dr. Hong reentered the car, flustered by the now more than 15-minute delay. Eventually, we were waved through the tunnel and continued on our way to Pyongyang.

On our final afternoon, we made a surprise visit to the new campus of Pyongyang University of Science and Technology (PUST), a visit that our group had requested but was declined for reasons never explained. Due to open the following year,

PUST was conceived and funded by Dr. James Chin Kyung Kim, a Korean-born American businessman from Florida. Financial contributions from Evangelical Christian groups established PUST as the first and only English-language university in North Korea.[20]

Back at the hotel, our consortium met to review our cooperative statement. We proposed an effort to build on Syracuse University's work and provide digital access to scientific research, as well as programs for teaching English to Koreans, new scientific exchanges, and a variety of environmental studies. We couldn't provide dollar figures for financing the projects, as we had no authority to discuss funding. Hong said it would take the government a month to review our research plan.

Our final dinner was scheduled to take place in the revolving platform attached to the top of our 44-story hotel. The US consortium sponsored and organized the dinner from menus not available elsewhere in the hotel. As we sat around a large table with our North Korean hosts and other senior DPRK officials, I made a final toast and described how to win a Nobel Prize. I was wearing a navy blue Johns Hopkins necktie, the same tie that I wore when I presented my Nobel lecture in Stockholm. On the back of the tie, I had listed the names of our US consortium members, and as I untied it from around my neck, I explained the significance of the escutcheon and inscription bearing Sir William Osler's motto, *Aequanimitas*. "It means presence of mind and clearness of judgment in moments of grave peril," I said. It was also the title of Osler's farewell address to graduating medical students at the University of Pennsylvania School of Medicine in 1889. It summed up my beliefs that scientists—and scientific diplomats—needed to think clearly and calmly at the most difficult times.

I presented the necktie to Dr. Ri Song-uk, vice president of the State Academy of Science and requested that he present the necktie to the first of his countrymen to win a Nobel Prize. Perhaps it was a tad melodramatic, but our wishes were sincere, and the gesture signaled the end of a valuable trip. Dr. Ri responded with a big smile and nodded his approval of the gesture.

As we prepared to leave, a smiling James Kim, the chancellor of PUST, arrived. Since our meeting with Chinese diplomats in Beijing, I had waited for a North Korean official or senior academic to say the words that would signal one metric of our success: a return visit. And as we shook Kim's hand, he uttered just those words: we were invited back to PUST.

But future collaborations would not be straightforward. We would learn from our experiences, and those of other scientists, that navigating science and sanctions could mean that political forces abruptly alter the best-laid plans.

FOLLOW-UP MEETINGS WITH SAOS
IN THE US AND ITALY

Fourteen months later, an SAOS delegation of North Korean scientists met with the US-DPRK Science Engagement Consortium at the Carter Center in Atlanta, Georgia, to continue the work begun in Pyongyang. Alongside Dr. Hong were Kang Sam-hyon (who became the DPRK's ambassador to Iran) and Ri Yong-gil (who became deputy head of the DPRK mission to the United Nations). We had planned to hold the meeting in Washington, DC, but the US Department of State rejected the idea and approved a meeting in Atlanta instead. The Carter Center was happy to host our delegation; President Jimmy Carter was highly respected in North Korea for his 1994 trip to Pyongyang to meet with Kim Il-sung and provide advice about how to protect people from a horrific famine that eventually left at least 2.5 million dead.

Vaughan Turekian said this meeting was used to show that partnerships are about much more than finances. "One of the key issues here was to demonstrate that cooperation isn't about exchanging money or equipment, but rather having follow-up discussions to better understand the systems in the two countries and how they operate," he said. But the February 2011 meeting did not begin well. After what we thought had been a successful trip to the DPRK in 2009, Dr. Hong opened by reading a DPRK government statement that called our visit a "waste of time," words that made me sit up straight. He went on to read a long list of serious complaints from his government that outlined their disappointment at the fact that no financial resources had been provided since our visit.

With no heads-up so that we could plan a response, I chose to tell a story instead. Before we left the DPRK in 2009, I had handed Dr. Hong an unopened carton of granola bars for his grandson, the child who had warned his grandfather to carry a rifle when dealing with Americans. I asked Dr. Hong, "Did your grandson tell you to bring the rifle to the Carter Center here in Atlanta?" "No," said Dr. Hong with a smile. "He asked me to bring back American candy!" My attempt to shift the conversation by talking about his grandson lightened the atmosphere, but it was clear that the consortium's work was under constant political pressure from DPRK leadership. No cooperation would occur without a large transfer of funds from the US government to Pyongyang—something that was not going to happen.

During the three-day visit in Atlanta, our group tried to steer the focus back to our statement of agreement with SAOS in 2009. We agreed that the first two priorities should be English-language training for North Korean scientists and the launch of virtual science libraries. We planned to speak monthly through the DPRK Mission to the United Nations. Our group had expanded to include faculty from Georgia Tech, the University of Georgia, Emory University, the University of Missouri, Stanford University, and officers from the Carnegie Endowment.

A second follow-up meeting of the US-DPRK Science Engagement Consortium with Dr. Hong and his SAOS colleagues was held in April 2012 at the Rockefeller Bellagio Center on Lake Como in Italy. That meeting went smoothly, with consensus on our work and a clear outline of the programs and research we hoped to continue. But progress was slowed by the heavy weight of sanctions. Political tensions remained ever-present at our meetings, even as we tried to focus on digital libraries and campus exchanges.

That was to be my final meeting with Dr. Hong, whom I had come to respect and admire. We exchanged gifts as we departed; Dr. Hong gave me a small wooden turtle. At a candy shop in the village, I found a small toy truck and a bag of candy. I hope that Dr. Hong's grandson enjoyed the present.

PYONGYANG UNIVERSITY OF SCIENCE AND TECHNOLOGY (PUST)

Pyongyang University of Science and Technology opened its doors to 50 male undergraduates in October 2010. Enrollment blossomed and eventually reached a steady state of approximately 600 undergrads and 100 graduate students, of whom less than 5 percent are women. The only English-speaking university in the DPRK, PUST provides a unique opportunity for Americans to meet and potentially bond with high-achieving

North Korean students who may become future leaders of the DPRK. Likewise, having full-time English language instruction may help the DPRK onto the global stage.

PUST was conceived, organized, funded, and led by its founding president, James Chin Kyung Kim, a Korean-born American businessman living in Florida. Financial contributions from Evangelical Christian groups from around the world yielded $45 million. Every PUST staff member is expected to adhere to all DPRK regulations, and religious proselytizing is absolutely forbidden. To operate within the DPRK, all leadership positions had dual appointees; each non-DPRK professor was matched with a DPRK counterpart.

The faculty are volunteers and teach at least one course per semester. Most faculty are members of Evangelical communities from the United States, South Korea, or Europe. I found the professors soft-spoken and diligent. Undoubtedly, many are excellent teachers. For them, teaching at PUST is a religious calling to spread the gospel through science education. Other faculty included young adventurers looking for unusual cultural experiences.

The PUST students are highly motivated and very bright. Most are children of the DPRK elite living in and near Pyongyang. Their experiences are narrow, and they don't question the ideological propaganda that tells them they are the luckiest people in the world and should never wish to leave North Korea. They have never heard otherwise. The great majority of PUST students expect to have careers involving computer software engineering and live comfortably in the DPRK. Social mixing of students and faculty is restricted. The students receive ideology instructions each evening while the PUST faculty gather for religious worship, but students are strictly forbidden to participate in or observe worship.

2011 PUST INTERNATIONAL CONFERENCE ON SCIENCE & ENVIRONMENT FOR WORLD PEACE

PUST held its first international science meeting on campus from October 4–8, 2011, with future symposia planned for every second year. My second journey to North Korea was during PUST's second year of teaching students.

As a DPRK requirement, all leadership positions had dual appointees; each non-DPRK professor was matched with a DPRK counterpart. President Kim's DPRK counterpart was Kwang-il Oh, a distinguished-looking gentleman about age 60.

I was impressed by several notable individuals who would attend the conference, called "PICoSEP." The PUST advisors included Malcolm Gillis, a renowned economist and former president of Rice University. Stuart Thorson from Syracuse University and Norman Neureiter from AAAS were the first scientific attachés assigned to the US embassies in Warsaw and Moscow during the Nixon administration. Norm was also the first science advisor to a US Secretary of State, Colin Powell, during the George W. Bush administration. Unfortunately, Hans Jornvall, a former secretary general of the Nobel Committee, for whom I had requested an invitation, was denied a visa by DPRK authorities without explanation.

Wearing black suits and monochromatic neckties, the all-male student body marched in formation, singing patriotic songs, as they arrived at the conference building. Norm and I were seated in the front row next to the other keynote speaker, Lord David Alton from Liverpool University. At the rear were the 200 PUST undergrad and 46 grad students, all sitting bolt-upright and ready to listen.

The program began with formal welcome speeches by President Kim and Vice Minister Jon. Then I was introduced. To combat the stiffness of formal keynote lectures, I often begin with a joke to determine whether the audience can hear and understand my presentation. After I told a silly story about a Nobel laureate, the international visitors chuckled. After a pause of a few seconds, the PUST students laughed in unison, apparently waiting for a cue. Afterward, we were taken on a tour of the 240-acre campus. The classrooms were well furnished and comfortable; all contained paired photos of Kim Il-sung and Kim Jong-un. Regrettably, the laboratories had a dearth of basic instrumentation such as pH meters and centrifuges—explained by severe budget constraints and the international sanctions blocking the import of "dual-use" equipment.

Afternoon lectures were given by the impressive international visitors on a range of topics. Former NASA astronaut David Hilmers spoke about his four missions to the International Space Station. Norman Neureiter spoke of his experiences in international science and technology cooperation. Randy Giles, president of the Bell Labs in Seoul, spoke. The Syracuse–Kim Chaek partnership was featured in the lecture by our colleague Stuart Thorson. A dozen presentations by DPRK faculty from Kim Chaek University and Kim Il-sung University ranged from economic modeling to the removal of environmental toxins.

Three PUST grad students sat with me at dinner and impressed me with their articulate spoken English and courteous manners. I gave them each my business card, which two seemed glad to receive, but the third declined politely, stating that he had no interest outside of the DPRK. They seemed very shy and asked few questions, so I asked them about themselves. All three were raised in Pyongyang by families with significant positions in medicine or government agencies, yet none had visited their families for the past year. Each expected to pursue a career in computer engineering within the DPRK. None asked me any questions about my career in the United States.

The Evangelical Christian faith of many faculty was never apparent when students were present, but in private discussions, it was less cryptic. Evangelical Christian faculty members were allowed to hold a private meeting every evening without interruptions, but PUST students were not allowed to attend. DPRK faculty and students attended obligatory son-gun (army first) and juche (self-reliance) education sessions in the evening, but non-DPRK persons were forbidden to attend. Although rigid, the system was perceived as satisfactory by both groups.

On the second day of the two-day conference, concurrent sessions were organized by topic: information technology, life sciences, management, and environment and collaboration. DPRK scientists presented lectures on various technical applications, emphasizing theory and mathematical equations. During the Q&A sessions, the PUST students posed formal, carefully scripted questions to the speakers.

President Kim, in his exuberant style, took me to visit the classrooms. The students were very friendly with President Kim, and it was clear that he had established a strong bond with them. They seemed eager to have their photos taken and appeared much more relaxed around the international visitors as we got to know each other.

After lunch, President Kim and Chancellor Chan-Mo Park took me to their office to discuss the end-of-conference farewell speech, normally given by a senior faculty member. They wanted me to read the farewell speech and had drafted a text. I scanned the two pages and replied that I found it repeated flowery phrases and other egregious exaggerations. It thanked the Great Leader for personally organizing the PICoSEP meeting and making PUST a world-class research university. They said I could edit the text a little, but they were required to show it to the DPRK superiors in the Ministry of Education. I edited the text, including a few of the most obsequious statements. When I showed the text to Norm Neureiter, he felt that I should read it because it could increase the support from the DPRK leadership in the DPRK Education Ministry. And so I did.

That afternoon, the PICoSEP conference ended with a closing ceremony with Vice Minister Jon Kuk-man, his Ministry of Education staff, and the entire PUST community. An interpreter stood on the podium for the non-English speakers. Near the conclusion, I was asked to present the farewell speech, and I slowly read it aloud, pausing for the interpreter. At the instant I stopped reading, the students sprang to their feet in a vigorous standing ovation. But as the applause died down, the interpreter looked concerned that I did not leave the podium. He looked alarmed when I spoke without a script.

> I wish to add my own heartfelt best wishes to the students of Pyongyang University of Science and Technology. We welcome you to the global community of science.

Afterward, the international visitors applauded. While it was a minimal gesture, I felt it was important to be candid. All agreed that the 2011 PICoSEP meeting was a success.

After the dinner reception, I returned to my apartment. The weather remained very warm as I walked past the high-rise student dorms in the early evening. When I had walked by earlier in the day, the undergraduate students in their black suits and ties were in their rooms studying at their desks in front of the large windows. But the heat apparently made them uncomfortable, so they all sat in their T-shirts and boxer shorts in front of the large windows, silently looking down and watching me as I passed. I couldn't resist teasing them and called out, "Where are your suits?" They must have found this amusing as peals of laughter rang out across the courtyard. Perhaps the students had relaxed some of their artificial formality.

2015 INTERNATIONAL SCIENCE MEETING AT PUST

Norman Neureiter and I made our third trip to North Korea in October 2015 to attend PICoSEP once again. A school for medicine and dentistry opened in 2014 but without a teaching hospital on campus. Skepticism about PUST surviving had proved unfounded despite the threat that the DPRK leadership could close the university down at any time. Its first class of students would soon graduate.

Changes were notable since my first trip to the DPRK six years earlier. Pyongyang had undergone an increase in population, and new construction was evident. Kim Jong-il had been dead for four years, and posters showed the smiling faces of Kim Il-sung and Kim Jong-il, this time together. The roadside posters depicting grotesquely configured US soldiers being bayonetted by heroic North Korean marines that we saw previously had all disappeared. Large, brightly illuminated river boats were tethered on the Taedong River with multiple upscale restaurants on each. More automobiles and even a few taxis were on the road. A friendlier atmosphere prevailed. We were lodged in Pyongyang at the tall Yanggakdo International Hotel, with excellent views of the city. Because the hotel is on an island, we were permitted to stroll outside without chaperones.

During our sightseeing, we could observe—but not meet—a preselected subset of the North Korean people. In the city, we encountered several groups of North Korean preadolescent school children dressed in colorful sports warm-up suits; the girls smiled at us and giggled while the boys wore ball caps on backward and seemed oblivious. Our hosts did a fine job of showing us interesting and attractive sites, including the lovely Mount Myohyang National Park a hundred miles northeast of Pyongyang, where we enjoyed a delicious outdoor Korean-style barbecue. Attractive young North Korean women in traditional gowns (*hanbok*) served as guides and hostesses to our group of foreign visitors. Another side trip brought us to the Taedonggang Fruit Farm, a beautiful valley filled with apple trees—the product of SAOS agricultural research.

Viewed superficially, one might conclude that the DPRK is indeed a lovely place, but the tours, of course, did not bring us anywhere near the impoverished villages and labor camps where life is an ordeal of deficient nutrition and political savagery further worsened by brutal living conditions and fierce weather.

VOLCANIC DISRUPTIONS AND SCIENCE UNDER SANCTIONS

The volcano was lifting higher into the atmosphere, its peak shifting toward the heavens by a centimeter each year while earthquakes rattled its base. But this was not just any active volcano. Mount Baekdu's major eruption cited as the year 946 CE was one of the most powerful volcanic eruptions in human history, ejecting upwards of 25 cubic miles of rocks, showering ash over more than half a million square miles, and causing a period of regional climate change. Mount Baekdu also had had at least three smaller eruptions, the last one in 1903.

Mount Baekdu would provide an opportunity for a crucial collaboration between scientists from universities and agencies in the United Kingdom, the United States, China, and North Korea, all working together inside the DPRK to make vital discoveries.

Volcanologists quickly extinguished any presumption that those toxic ash clouds and devastating magma had spilled more than a millennium ago, in 946 CE. "That happened very,

very recently," says James Hammond, a British volcano scientist.[21] Hammond speaks in geological terms where a volcanic eruption thousands of years ago seems like "yesterday"—and poses a threat to humanity today.

Mount Baekdu (Whitehead Mountain) is a significant site for Koreans. Straddling the border between China and North Korea, the 9,003-foot volcano is considered the spiritual home of North and South Korea and features in both countries' national anthems. The highest point on the Korean Peninsula, the volcano is mythologized as the birthplace of Dangun, the founder of Gojoseon, Korea's first kingdom. It's been venerated by Korean kingdoms ever since.[22] Its crater-top lake and snow-capped peak sit proudly atop the national emblem of the Democratic People's Republic of Korea and are emblazoned across ornaments and flags.

So, when Mount Baekdu woke from a century-long rest in 2000, with magma from its depths shifting upward, its earthquakes increasing by two orders of magnitude, and an efflux of gases escaping its crater, there were fears that it could erupt as dramatically as the explosion of 946. Increased volcanic activity can be a sign of imminent eruption, although researchers are still unsure when or why volcanic unrest sometimes leads to an eruption and sometimes simmers down to another protracted geological snooze.

Mount Baekdu's history of devastation led North Korean researchers to seek help from outside experts. But not immediately. Local scientists monitored the situation from 2000 to 2006 while the volcano was in a state of unrest; their work garnered international attention, including news reports and conversations at scientific meetings. They first sought outside expertise from China. Then, in 2011, James Hammond, the British seismologist (Imperial College), and his colleague Clive Oppenheimer (University of Cambridge) were invited to North Korea to journey to the volcano's peak. North Korean scientists didn't frame the invitation as a request for help, but rather as an interesting opportunity to collaborate.

Hammond was stunned. He and Oppenheimer would be among very few non–North Korean scientists ever to gain access to the sacred volcano. But first, there was the issue of travel to a nation burdened with sanctions and embroiled in tricky diplomatic relations. While the AAAS and the Richard Lounsbery Foundation, a nonprofit invested in strengthening America's science networks through international collaboration, offered financial support and guidance for the historic trip, Hammond could not get his employer to sign off on his travel. He was 29 years old, not long out of graduate school, and not yet a permanent employee at Imperial College London. Still, he had valuable experience planning research in another sanctioned region, the border between Ethiopia and Eritrea.

Hammond was a brand-new research fellow, and the first piece of paperwork he asked the university to sign was a memorandum of understanding with the DPRK. Hammond reassured administrators that there would be no cash flow from North Korea to the university. Still, university officials were nervous about co-signing a document of agreement with a sanctioned nation that boasted of its nuclear prowess.

In September 2011, Hammond and Oppenheimer arrived in Pyongyang. Unlike the United States, the United Kingdom maintained an embassy and full diplomatic relations with the DPRK, and Hammond and Oppenheimer were both UK citizens. Little did Hammond realize that this was to be his first of eight trips to the DPRK.

They woke the next morning to a national parade on the streets outside their hotel, the anniversary celebration of the Korean Workers' Party. They also received a warm welcome from a group of North Korean scientists. "We had no idea what was expected of us," said Hammond. "We knew we would arrive and fly up to the volcano, but that was all we knew."[23] Hammond showed his peers the seismometer he had brought from London, which he would use atop the volcano to measure movements in the ground.

A chartered plane took Hammond and Oppenheimer to the base, from where they trekked to observation stations positioned thousands of feet high on Mount Baekdu. Here, they met with 30 North Korean scientists, including geologists and geophysicists. Over four days, the scientists sampled the lake in the volcano's crater, collected gas specimens, and recorded seismometer readings. They learned that molten rock from the Earth's mantle, more than 20 miles beneath their feet, had soared to a much shallower depth, only three miles beneath the volcano, resulting in six years of volcanic unrest.[24]

Some trickier elements of international collaborations include developing scientific research questions, securing funding, and assigning team responsibilities. But on Mount Baekdu, the Western and North Korean scientists quickly agreed on two key research questions: What is the volcano's current state of activity, and what is its history of eruptions?

After four days of talks and observations, Hammond and Oppenheimer flew back to England full of optimism. Research questions established: check. Team responsibilities assigned: check. Funding secured: check. But months later, the smooth and amicable start hit rocky terrain.

"We were naive," said Hammond. "We were just scientists talking to scientists with no real understanding of the geopolitics or what you'd read in the newspaper." Soon after he returned to England, the DPRK planned ballistic missile launches, military exercises, and nuclear tests. British and American sanctions followed, complicating the team's clear-cut research plan.

The group's success in studying the volcano would impact the lives of many more people than those residing in the shadows of Mount Baekdu. One in ten humans lives within 60 miles of a volcano with the potential to erupt.[25] But millions of these people live in countries with international sanctions, including Syria, Sudan, Russia, the Democratic Republic of the Congo, and the DPRK. Together, sanctioned nations are home to 193 volcanoes.[26] Strained relations between these countries and the rest of the world restrict volcanic research, which can delay and impede disaster preparedness. There remains a lack of understanding about the inner workings of volcanoes: Why do some erupt, others smolder, and some remain asleep for centuries?

The effects of an eruption can spread across continents. In fact, volcanologists are certain that Mount Baekdu's major eruption occurred in the winter of 946 because of ash deposits discovered in deep ice layers more than 8,000 miles away in Antarctica. Studying the icy depths at which the ash was found allowed researchers to date the event.

Iceland's Eyjafjallajökull volcanic eruption in 2010 was relatively small compared with the explosion of Mount Baekdu in 946, but its effects were felt across Europe. Air traffic was disrupted for six days, and the cost to commerce and travel was estimated at $5 billion.[27] Beyond flight delays and canceled holidays, volcanic eruptions can accelerate climate change. When Indonesia's Mount Tambora erupted in April 1815, Earth's temperature fell by nearly 40 degrees Fahrenheit, causing failed crops and famine across parts of North America and Europe and what came to be known as "the year without a summer."[28]

Hammond and Oppenheimer's invitation to collaborate with North Korean scientists was a rare and valuable opportunity to expand scientific understanding of volcanoes and take steps to protect millions of lives—and planetary health. But politics stood in the way. To conduct science under sanctions and to navigate research in a politically strained environment, Hammond and Oppenheimer turned to lessons they had learned while studying a different geological phenomenon in another sanctioned nation more than five thousand miles away.

The pair distilled their success in Eritrea into five key lessons to be applied to scientific collaborations in other politically strained regions:

1. Establish clear research goals that are obviously distinct from political agendas;
2. Develop modes of communication that are clear and transparent;
3. Choose enthusiastic collaborators wherever possible;
4. Balance scientific ambition with real-world pragmatism so you can deliver on promises and show incremental success; and
5. Remain flexible in the face of unexpected barriers.

They witnessed the importance of developing networks of allies—people in government, academia, and industry who are invested in the success of cross-border scientific collaborations. So, when political constraints complicated their research in the DPRK, they drew up a network of allies who could facilitate volcanology research, even as missiles were being tested and new sanctions were being imposed.

Then, another hurdle emerged. The team had established that clear communication was one of the tenets central to successful cross-border work, but once Hammond and Oppenheimer were home in England, they couldn't contact the North Korean geologists they had met during their research trip. Hammond reached out to nongovernmental organizations based in the DPRK that had decades of experience working with Korean officials and asked them to serve as conduits. Essential organizations included the Pyongyang International Information Center of New Technology and Economy, the Earthquake Research Center in Pyongyang, and the Environmental Education Media Project in Beijing.

Using those channels, they arranged in-person meetings for Korean, American, and British scientists in France, Germany, and China. "We lacked trust early on just simply because we didn't know each other," said Hammond, reflecting on some of those early meetings in Europe.[29] "A lot of what we talked about was done through translators, and when you're talking about very technical topics, getting that information across can be difficult. But that got better over time as we got to know one another."[30] This is where enthusiasm, another lesson gleaned during the Eritrean project, came in handy. Beset by political constraints and hostile governments, the team of international scientists remained motivated to make the collaboration work.

Then there was another hurdle. Export licenses are required to comply with sanction regulations, and they require careful negotiations and handling. Hammond credits the Royal Society and the AAAS Center for Science Diplomacy for expert advice in this domain. But once export licenses for the equipment were secured, the team faced another challenge. Monitoring volcanic activity requires detecting where molten rock lies beneath a volcano, and it relies on equipment called magnetotelluric machines to read those electromagnetic changes. UK government officials told Hammond that the same machines could be used to detect submarines, making them "dual-use" items with the potential for exploitation by foreign militaries.

Equipment necessary for the team's research fell squarely on the contraband list. Still, Hammond remained hopeful. For two years, he waited as British officials talked with American officials who talked with Korean officials. But to no avail. With lost time and sidelined equipment, the group went back to the drawing board. They decided to use regular seismometers, which were not contraband, and to deploy them for longer periods in the hopes of gathering the right data.

This was not science for the rigid, unimaginative, or impatient. Remaining nimble, creative, and optimistic as hurdle after hurdle appeared was central to the group's success.

"We explained the situation to our North Korean colleagues with complete openness and honesty," said Hammond about the machines that could be used for military purposes. "It must have been frustrating from their perspective, but they understood that we have to work within international regulations."[31]

In August 2013, Hammond and Oppenheimer traveled to the DPRK, not with the ideal machines, but with seismometers, and got to work (chronicled by Richard Stone). North Korean scientists created the optimal placements for those seismometers, and data was steadily collected. The only woman on the team was an American, Kayla Iacovino, a volcanologist representing the US Geological Survey. In 2016, the team would publish two high-impact research articles coauthored by North Korean, Chinese, UK, and US scientists, and they jointly presented their work at international symposia. They hope their work will help uncover some of the mysteries of volcanic interruptions, information that might eventually protect the millions of people, including North Koreans, who live in the shadow of volcanoes. Their publications set very high standards for efforts involving international scientific collaboration.

Hammond, who is now a lecturer in geophysics in the Department of Earth and Planetary Sciences at Birkbeck, University of London, believes that these successes spurred new ambitions. "As we deliver our promises, such as having research visits to London and jointly publishing papers together, it means we can establish trust and that means we're doing much more ambitious projects now than we were initially."[32]

But even as articles were published and presentations delivered, obstacles continued to emerge. In 2015, the DPRK offered to discontinue nuclear missile testing in exchange for the suspension of joint US and Republic of Korea military exercises. The United States rejected this offer, leading to heightened tensions culminating in a new round of missile testing by the DPRK. These instigated renewed cycles of testing and harsher sanctions. A consortium of American and DPRK scientists that had planned a series of scientific workshops was told that sanctions prohibited international research, and the meetings were canceled.

What began as limited but exciting opportunities for research have narrowed over time. Still, Hammond remains focused on all that has been achieved. "I'm most excited about how long-lasting and persistent the collaboration has been," he said, listing leadership changes in the DPRK, the United Kingdom, and the United States, and military exercises, nuclear tests, and ballistic missile launches that have slowed essential research on volcanoes. "It's been remarkable that we've transcended some of that political turmoil."[33]

Mount Baekdu did not erupt at the beginning of the twenty-first century, but scientists are none the wiser about its return to a state of rest. "It didn't erupt, but it could have," said Hammond. "And we are yet to understand why it didn't or when it might."[34]

CURRENT CHALLENGES FACING THE PEOPLE OF THE DPRK

Volcanoes are not the only threat to North Koreans. Nearly half the population of North Korea is undernourished, according to the United Nations. Widespread stunting, the highest tuberculosis rates outside of sub-Saharan Africa, and epidemics of hepatitis B are some of its biggest current and looming public health disasters.[35] Because of its subpar health care system, the lowest health care spending of any nation in the world, a lack of disease-specific researchers, and emphasis on military and atomic funding, the DPRK will need the help of international scientists to counter these challenges.

Two different threats offer alternate paths to forging new scientific relationships: biodiversity and climate change. Around 40 percent of North Korea's forests have disappeared since 1985, according to data from satellite images.[36] Many forested areas have been converted into agricultural systems to supply food. Wildfires, infestations, landslides, and deforestation to supply firewood have also contributed to and increased the likelihood of further erosion, flooding, and biodiversity loss. While most of the Mississippi-sized nation has fertile soil, thanks to the volcanic terrain, this environmental degradation has made it difficult to cultivate arable land across North Korea.

A 2003 report by the UN Environment Program listed biodiversity, land degradation, water quality, forest depletion, and air pollution as five key environmental issues that needed to be urgently addressed in North Korea. After severe flooding in 2016, the country's Supreme Leader, Kim Jong-un, recognized his nation's vulnerability to climate change and declared forestry research and afforestation—planting trees in areas with no previous tree cover—urgent tasks.[37] When the country launched its 10-year plan for afforestation and reforestation, it listed science as the foundation for these regenerative efforts and the critical basis for addressing sustainability. Much of this expertise exists outside of North Korea, while scientists in the West rely on expertise from Asia to address new types of infestations damaging forests across the Western Hemisphere. Many forest-damaging pests around the world have originated in Asia.

In 2018, North and South Korea agreed to scientific collaboration in these areas, partly because existential threats do not stop at political borders or respect international sanctions. But progress has been slow and difficult to measure while climate change and biodiversity issues become increasingly urgent. A growing number of slow but successful cross-border partnerships, including the work of Hammond and Oppenheimer, offer hope that ongoing international collaboration with scientists in the West is finally a possibility.

THE UNRAVELING OF US-DPRK RELATIONS

Politics would hold hostage the American seismometers on Mount Baekdu's peak. Despite the best efforts of scientists collaborating across borders, autocrats' words would halt the advancement of science, but that wasn't known when the equipment was shipped to North Korea.

The Mount Baekdu collaboration and discussions of other potential collaborations between DPRK scientists and the AAAS suggested a new level of cooperation between our countries, but the optimistic tone changed quickly. Kim Jong-un's arrival at a time of profound national sadness over the death of his father, Kim Jong-il, coincided with a new wave of American relations with the DPRK that would turn in a negative direction.

The youthful Kim Jong-un was initially underestimated but soon exhibited ruthlessness to those he perceived as threats. His uncle, Jang Song-thaek, a close contact with leadership in China, was publicly humiliated and executed by machine gun in 2013. Prominent members of the DPRK leadership were swiftly removed for alleged lack of loyalty, and terror swept through the ranks as more than a hundred ranking government members, including the President of the State Academy of Sciences, were expelled and sent to so-called reeducation camps where many were executed.[38] Kim Jong-un's expatriated older brother, Kim Jong-nam, once considered the likely heir to Kim Jong-il, was assassinated with the VX nerve agent at Kuala Lumpur International Airport in 2017.

Then an American student from the University of Virginia was arrested in Pyongyang for a petty offense involving the theft of a propaganda poster.[39] Otto Warmbier, part of a student tour of North Korea, was sentenced to 15 years of hard labor in 2016 but was sent back to the United States in a vegetative state; he died days after his return. North Korean officials said Warmbier had contracted botulism and taken a sleeping pill; American coroners said he had suffered a head injury. Later that year, two Korean-American faculty at PUST, Kim Sang-duk and Kim Hak-song, were imprisoned for "hostile acts" against the DPRK.[40] Both were released in May 2018 after a visit to the DPRK by US Secretary of State Mike Pompeo.

President Donald Trump issued an order banning US citizens from visiting the DPRK in September 2017, two days before North Korea exploded its sixth and largest nuclear test. The 250 kiloton explosion was a thermonuclear device (a hydrogen bomb that released 20 times the energy of the Hiroshima atomic bomb) and should be taken "literally," stated Ri Yong-phil, a senior official in the DPRK Foreign Ministry. (Ri had been part of the DPRK delegation that we met with in Atlanta in February 2011.)

After preliminary summit meetings in Singapore in 2018 and Hanoi in 2019, Donald Trump, Kim Jong-un, and Moon Jae-in met at the 38th parallel. This prompted intense interest from the international news media. But despite the fanfare and statements of affection, no advance conditions were set, and talks broke down. The DPRK resumed the development and testing of missiles.[41]

Despite significant advances in scientific exchanges, the DPRK has at least temporarily returned to its former status as the Hermit Kingdom. The question remains: Will it be possible to resume scientific discussions?

Fifteen years after our first visit to the country, Vaughan Turekian is pragmatic about the progress—or lack of—that has occurred. "It is clear that any momentum to move forward ended with the political changes in the DPRK. There isn't really any follow-up taking place right now. Perhaps at some point, there will be a desire to move toward a more stable relationship, and communications between and among technical experts can help support dialogue. But that currently seems far off."

CHAPTER 4

Sub-Saharan Africa
INFECTIOUS AGENTS OF CHANGE

IN THE WINTER OF 2010, Egyptian and German doctors and scientists worked with the Egyptian government's Supreme Council of Antiquities to solve a 3,300-year-old mystery: What killed the golden boy of ancient Egypt, the teenaged pharaoh, King Tutankhamun? Genetic analysis and CT scanning of the mummy revealed that the young king's remains were riddled with the genetic leftovers of not one but multiple strains of the parasites that cause malaria. As scientists sifted through the teenager's tomb, they found the DNA of the most vicious malaria-causing parasite, *Plasmodium falciparum*.[1] Finally, here was clinical proof. King Tutankhamun suffered multiple bouts of malaria throughout his short life; the disease killed him in 1324 BCE when he was just 19 years old.[2]

This discovery—the first time Egyptian officials had allowed geneticists to probe royal mummies—offered the oldest proof of malaria infection. The researchers were quick to point out that it was no surprise the pharaoh had suffered malaria; more than half of all humans who have ever lived were most likely infected with malaria.[3] Unfortunately, malaria continues to bewilder scientists and remains a major disabler of adults and a leading killer of children. Even as technological advances soared early in the twenty-first century, malaria continues its death march, in 2022 killing 619,000 people—most of them African children who will not live to see their sixth birthdays[4]—and causing 247 million people to suffer illness.[5]

Malaria is caused by the microscopic *Plasmodium* parasite, transmitted to humans during a bite from an infected mosquito. The parasite infects red blood cells and causes pain, fevers, organ damage, and death. Malaria has had a profound influence on human evolution. So pervasive and loathsome is malaria that human DNA has accumulated mutations that protect against malaria's crippling fevers and wrenching pains. One such mutation is found in the gene encoding beta-globin, a component of hemoglobin, which helps blood cells carry oxygen. Normally, red blood cells carry oxygen complexed to hemoglobin-A, a protein formed from two alpha-globin chains and two beta-globin chains (referred to as Hb AA).[6] But patients with sickle cell disease (Hb SS) inherit two mutant genes, one from each parent. These mutations cause sickle cell disease, in which red blood cells stretch from their regular donut shapes into jagged crescents that jam inside blood vessels, cutting off the oxygen supply to organs and tissues and causing excruciating pain.

But people who inherit just one mutant gene are said to have sickle cell trait, a harmless condition in which only a few blood cells form sickles. In contrast to people with sickle cell *disease* who die in relatively young, people with sickle cell *trait* can live long and healthy lives. In fact, those with sickle cell trait enjoy relative resistance to malaria, giving them an above-average survival rate compared to others living in malaria-endemic regions. So great is the advantage of having sickle cell trait that 10 to 40 percent of people across equatorial Africa have inherited one mutant gene from a parent.

Malaria has changed our DNA and the way we plan families. In sub-Saharan Africa, which has the highest malaria burden in the world, people often give birth to more children than in other parts of the continent, expressly to make up for those children who will die from malaria. Increasing the size of families beyond the malaria death rate may leave parents with more mouths to feed than they can afford.[7]

Malaria has also shaped the global public health map, dictating where we build public health organizations. The Centers for Disease Control and Prevention (CDC), an agency of the US government, is based not in Washington, DC, like so many federal organizations, but in Atlanta, because of malaria. In the early twentieth century, the American South was blighted by parasitic diseases, including endemic malaria. The CDC is the direct descendent of the Office of Malaria Control in War Areas, an agency tasked with protecting troops from an enemy pathogen during World War II.[8]

Global malaria killed between 150 and 300 million people in the twentieth century alone, accounting for 2 to 5 percent of all deaths. Today, the parasite impoverishes nations, costing Africa at least $12 billion in lost revenue each year,[9] and erasing 70 percent of the per-capita income of malaria-endemic countries.[10] In the twenty-first century, some countries spend 40 percent of their public health budgets on fighting malaria.[11]

Malaria has inspired public health efforts of gargantuan magnitudes and even toxic and authoritarian designs, sometimes requiring public health doctors to partner with regimes and militaries to fight the disease-causing parasites and the mosquitoes that spread them. American epidemiologist Fred Soper had an illustrious career with the International Health Board of the Rockefeller Foundation. Taking on the mantle of William Gorgas, the renowned US Army physician who, a generation earlier, had defeated yellow fever and malaria epidemics in Havana and at the Panama Canal by mosquito control, Soper emerged a formidable mosquito-fighter.[12] Using toxic insecticide sprays, Soper showed that mosquito control could provide remarkable benefits.

In 1927, Soper relocated to Brazil, where a yellow fever outbreak was raging. The aggressive daytime feeding habits of the *Aedes aegypti* mosquitoes posed a challenge. Often referred to as the "yellow fever mosquito," *Aedes aegypti* mosquitoes are also vectors for other viral pathogens: Dengue Fever Virus, Zika Virus, Chikungunya, and West Nile Virus.[13] *Aedes aegypti* is the most widespread of the almost two hundred *Aedes* mosquito species worldwide.[14] Aggressive feeders, *Aedes* mosquitoes repeatedly bite the unlucky human host.

Soper's approach to the yellow fever outbreak was rigorous bordering on dictatorial.[15] A meticulous planner, Soper commanded an army of inspectors who worked on a minute-by-minute schedule across a precisely carved-up map of each sector of town to check in every domicile the toilets, gutters, puddles, bathtubs—any place mosquitoes could breed—and spray infested homes with insecticides and treat outdoor breeding sites with larvicides. A supervisor checked each inspector's work, and penalties were issued for failure. Soper himself lost 20 pounds through his own physical contributions to this work. Eventually, the yellow fever outbreak subsided.

Malaria mosquitoes belong to the genus *Anopheles*. Of the four hundred species of *Anopheles* mosquitoes, only about 60 are known to transmit malaria to humans. The females of the night-feeding *Anopheles* species transmit the malaria parasites to humans, leaving a trail of feverish suffering in their wake. None is deadlier than the female *Anopheles gambiae* mosquito,[16] the dominant African mosquito, which transmits a tiny number of malaria sporozoites from its salivary glands into the skin of a sleeping human host. Within a few days, that host is stricken with a horrific febrile illness caused by billions of malaria parasites circulating through the bloodstream, damaging the brain, lungs, and other organs.

So, in 1938, when the worst outbreak of malaria in the Americas killed more than a hundred thousand people in northeast Brazil, Soper went even bigger. He had warned government officials eight years earlier that this outbreak was imminent and that stopping the mosquito larvae before maturation was imperative.[17] The event that fueled Soper's concern was a colleague's discovery of thousands of *Anopheles gambiae* larvae in a puddle near a port on the Brazilian coast. Soper concluded that the insects must have arrived in ships from Africa. Still, the Brazilian government did not heed his warnings. With an army of four thousand uniformed inspectors at his command, Soper dictated an even more ambitious mosquito eradication program. He declared war on the proliferating *Anopheles gambiae* mosquitoes over nearly twenty thousand square miles. It was a ludicrous goal. Soper aimed to wipe *Anopheles gambiae* off Brazil's shores, although the mosquito had already spread from its first sighting near the port along hundreds of miles of Brazilian coastline. It had even made strides inland, where *Anopheles darlingi* and other species of *Anopheles* mosquitoes were already established in their native habitat along the Amazon and throughout the interior. With military precision, Soper earned and re-earned his nickname, "The Commander,"[18] as the anti-mosquito army entered every home in each precisely mapped sector. Emboldened by government-issued powers and armed with toxic chemicals, they inspected, they doused, and they inspected again.

Perhaps this trifecta of field epidemiology, military prowess, and government authority was the only way to make progress in public health. While some eschewed the influx of strict governance into the scientific method, Soper's strategy successfully wiped out the mosquito—at least temporarily. In less than two years, he had achieved the impossible. *Anopheles gambiae* was nowhere to be found over the entire land mass. The Commander and his chemically armed troops had surveyed and sprayed. Certainly, Soper's method required toxic compounds and injunctions that forced people to open their doors to uniformed men who inspected and doused every crevice of their homes. Historians now believe that the 1938 malaria outbreak in northeast Brazil was much more complicated and that Soper overstated his achievements. Nevertheless, he was celebrated in his day.

Elected in 1947 as president of the Pan American Sanitary Bureau (predecessor to the Pan American Health Organization),[19] Soper championed operations around the globe to eradicate vector mosquitoes. In the words of William Gorgas, Soper's predecessor, who had himself cleared the Panama Canal of malaria at the turn of the twentieth century and was knighted on his deathbed for doing so: "Destroying mosquito larvae is essential; everything else is secondary to it." A unique figure in the history of global public health, Fred Soper shared one of the first Lasker Awards in 1946 "for his splendid organization of eradication campaigns against yellow fever and malaria, which have set new standards in the fight to defeat these diseases."

The use of toxic insecticide spraying declined abruptly in 1962 when Rachel Carson published her best-selling book *Silent*

Spring, signaling the launch of the environmental movement. Carson, a 1932 graduate of Johns Hopkins University, considered herself a conservationist; her writings ultimately led to a global ban on DDT use.[20] Mosquito control became more difficult, and global control of vector-borne diseases declined. Malaria, which had been eliminated in the island states of Sri Lanka and Zanzibar, reappeared after spray campaigns could no longer use DDT.

Spraying the countryside with toxic insecticide was not an approach that Philip Thuma was willing to take to end his patients' suffering in rural Zambia. A white man working in southern Africa, the son of a medical missionary,[21] Thuma was born in Pennsylvania in 1950 and raised in Zambia (formerly Northern Rhodesia), where his parents arrived when Thuma and his brother were young boys. The soft-spoken pediatrician had studied at a boarding school in Zimbabwe (formerly Southern Rhodesia), attended medical school in Pennsylvania, and completed four years of pediatrics residency plus one year as the chief resident at Johns Hopkins Children's Center. Like his father, the younger Thuma spoke the Chitonga language and deeply respected the Batonga people and their hierarchies and traditions.

Thuma's father, Alvan, was a legend in Zambia. The elder physician, a general practitioner from Ohio, had delivered babies, cradled the dying, and lent his name to dozens of newborn members of the Zambian Batonga tribe. In 1957, he built

Macha Mission Hospital in a town of the same name in Southern Province. The Macha Mission is affiliated with the Brethren in Christ Church (BIC). Related to the Mennonites, BIC has a strong presence in Zimbabwe and Zambia due to early twentieth-century missionaries.[22]

Macha was 40 miles from the closest paved road in the rural outback, between Victoria Falls and Lusaka. The average lifespan of Alvan's patients was only 44 years, and more than 10 percent of children died in their first year of life. Throughout his years working in the region, he witnessed hundreds of children die from malaria and cared for many more with epilepsy, learning difficulties, and deafness—lifelong disabilities inflicted by the malaria parasite.

Following in his father's footsteps, Thuma turned down the academic physician life offered to him at the end of his residency training in the United States, and he returned to Zambia with his wife and young children. He was intent on fighting one of Zambia's deadliest scourges. But, on his return, there was little he could do for the scores of young children who died from malaria during every rainy season. "I have seen kids under five who are looking fairly normal one day, fever and headache the next day, by the third day they're in bed, and by the fourth day they might be dead," Thuma often stressed during our many conversations at Macha, his beloved home in Zambia.

For over a century, the fight against malaria had split the anti-malaria brigade into two camps: Those like Gorgas and

Soper, who believed that eradicating mosquitoes with toxic insecticide spraying would end the epidemic, and those who believed that strategies targeting the parasite, such as medicines to treat the infected human host, were the answer. Thuma considered a multipronged, collaborative approach, one that relied on field epidemiology and boots-on-the-ground public health work. Thuma also knew he would need approval from Chief Macha and the local headmen. Financing health care in rural areas like Macha is a considerable expense for an under-resourced African nation like Zambia. The funds to build a research campus with modern laboratories, insectaries, offices, and residential accommodations for visiting researchers were simply unavailable.

Thuma's strong ties to the medical establishment in the United States and his relentless ambition to create a malaria research campus adjacent to Macha Mission Hospital drove him to submit a series of successful proposals to the United States Agency for International Development (USAID). The timing coincided with a unique opportunity to partner with the new Johns Hopkins Malaria Research Institute (JHMRI), which had received a magnificent gift from an anonymous donor—later disclosed to be Michael Bloomberg and the Bloomberg Philanthropies. In 2003, the new campus was christened the Macha Research Trust (MRT),[23] and a long-term partnership with JHMRI was formalized.

Just weeks after I became director of the JHMRI in 2008, I left my office in Baltimore to join Macha's field-testing teams as they worked to answer these questions. It would be the first of two or three trips I would make annually to Zambia over the next 12 years. That first morning, Harry Hamapumbu, the team's highly respected project field manager, arrived by bicycle at the MRT campus. I watched him swap two wheels for the MRT Land Cruiser and load it with supplies. I piled in with the rest of the team and wondered how they would navigate to villages and in the nearly roadless countryside.

Hamapumbu supplemented his own knowledge of the landscape with a specially designed GPS system, which could pinpoint the locations of the modest one-room dwellings in the countryside using data sent by cell phone text messages. As we trundled through the bush, it surprised me to see small solar recharging units in the outback and cell phone towers near villages with no running water or toilets. The system for texting health information to the Macha laboratories was first developed by Clive Shiff,[24] a Zimbabwe-borne parasitologist at Johns Hopkins. The Zambian government had been so impressed that they incorporated the technology into Zambia's national system.

Most Batonga people are subsistence farmers who cultivate maize and a few other crops by hand in small plots.[25] The region lacks any industries, and the only major employers are a nursing school and a single high school. As we drove from one village to another, speaking with farmers and their families, I listened as Hamapumbu seemed always to say the right words to explain to each community member exactly what kind of testing his team was doing and why. These visits often followed a recent case of malaria diagnosed by one of the dozen rural health clinics affiliated with the 208-bed Macha Mission Hospital. Asymptomatic family members and neighbors were tested to identify carriers of malaria—people suspected to be a source of infection for hungry *Anopheles* mosquitoes. Hamapumbu explained that anyone who tested positive for malaria parasites, even seemingly healthy individuals, would be offered anti-malaria medicines that could eventually rid them of the parasite and prevent more infections.

The traditional way to test for malaria includes smearing a drop of blood onto a glass slide. Under a microscope, malaria parasites can be seen inside red blood cells.[26] But if low levels of the parasites are present in a patient's blood, they can be easily missed by microscopy, which is impractical in remote villages during field visits. So, instead, Hamapumbu squeezed a drop of blood from a patient's finger onto a commercially available rapid diagnosis test (RDT) card. Fifteen minutes later, the RDT determined whether malaria parasites were present.

This active surveillance wasn't the only approach to reducing the suffering caused by malaria. In the early 2000s, Rebekah Kent was the first of a series of Johns Hopkins doctoral students to visit Macha and study the region's mosquitoes. She discovered 31 different species and identified the culprit behind the majority of Macha's malaria: The stream-loving mosquito

Anopheles arabiensis. Since mosquitoes can travel 1,500 feet to feed, Kent and colleagues used satellite imaging to map the area's hydrology to identify places where *Anopheles* mosquitoes put nearby residents at higher risk of malaria.

Field work in rural Africa is challenging due to isolation, extreme weather, harsh terrain, and the dangers of disease. Jenny Stevenson grew up in Kenya and graduated from Oxford and the London School of Hygiene & Tropical Medicine. She served as a MRT–Johns Hopkins entomologist in residence at Macha from 2013 to 2020. During her years at Macha, Jenny organized and directed the insectary and built the first mosquito house laboratory in Zambia on the MRT campus, where her protégé, Limonty Simubali, and a new generation of young Zambian scientists became trained in mosquito research.

Under Jenny's supervision, more than a dozen Johns Hopkins graduate students—nearly all women—studied epidemiology and entomology in the field at Macha and other rural sites in Zambia and Zimbabwe to ascertain which of the multiple *Anopheles* mosquito populations were responsible for indoor biting and which preferred outdoor attacks, requiring new control strategies.

When I asked Jenny whether she preferred the comforts of living in Geneva, where she serves as a Technical Officer at the World Health Organization (WHO), her answer was clear. "I miss life in Zambia and with JHMRI. Working with WHO has been a great learning experience, but I miss the field work and collaborating with people on the ground." When I asked her whether she was intimidated by the challenges of working in the field, she answered, "Not at all. That is where malaria is most devastating and where I feel we can accomplish the most. I anticipate returning to rural Africa in the not-too-distant future."

When Thuma began malaria research in the region, the doctors at Macha Mission Hospital used chloroquine and quinine to treat patients suffering from the disease. "In the days when we had quinine, we used to say that you're supposed to take seven days of it, but by day three, you were so dizzy, and your ears rang so much that you usually couldn't function, and you had to go to bed," Thuma told his alma mater's magazine. "In those early days, it was sometimes hard to separate out the symptoms of the disease from those of the drugs." Eventually, malaria parasites grew resistant to medicines designed to eliminate them, and Thuma became worried that the drugs were becoming useless. In the 1990s, he noted the mention of a new and promising malaria medicine in the pages of medical journals. The medication was first discovered in 1972 but was kept secret by the Chinese government and further neglected by pharmaceutical companies due to the presumed low profitability of a medicine for poverty-stricken malaria patients. Artemisinin, an extract of the herb sweet wormwood, became a miracle drug for treating malaria.

The discovery piqued Thuma's interest, and he organized a randomized, controlled trial of 92 Zambian children who were suffering from cerebral malaria to compare the effects of the standard treatment—the dizzying intravenous quinine for seven days—with the new drug, which was injected daily for five days.[27] Thuma had managed to get a small supply of the drug from a Dutch company that manufactured a derivative of artemisinin called artemotil or artemether. Thuma's trial showed no difference in survival, recovery time, or clearance of parasites between the old and new drugs. Still, the artemotil treatments did have one advantage: They were much easier to administer to patients and didn't cause the severely unpleasant side effects of the older medicines, although some children did experience neurological events.

This was enough to motivate Thuma and his team to continue studying artemisinin drugs. But in their next studies, they used a different version of the new malaria drug, dihydroartemisinin, which could be given by mouth, eliminating the need to inject patients with needles. Patients at the Macha Mission Hospital received either the new oral medicine or the standard treatment; their medical teams were intentionally kept unaware of who was receiving which treatment to reduce possible observer bias. Within weeks, a startling difference was apparent. Half of the patients in the study had entirely cleared the parasites from their blood within hours of starting therapy. Thuma was desperate to know which drugs these patients had received. He petitioned for the decoding of the study, and, as he had suspected, patients who had cleared the malaria parasites quickly had all been randomized to receive dihydroartemisin in.

Secretive Science

It was the war in Vietnam that led to the breakthrough in malaria treatment. During the 1960s and 1970s, US military personnel lost more time to malaria than to combat injuries. Regulars in the People's Army of North Vietnam were also incapacitated by chloroquine-resistant malaria. North Vietnam's prime minister, Ho Chi Minh, sought help from Chinese Premier Zhou Enlai.

In a secret meeting held in Beijing on May 23, 1967, Chairman Mao Zedong called for Chinese scientists to search for new medicines to fight malaria. Mao directed Chinese medicinal chemists to search for compounds and synthesize new mixtures that could work against malaria through an initiative code-named Project 523 (for the fifth month and twenty-third day). A few compounds were identified, including Lumefantrine, an agent that clears malaria parasites from the blood and remains active long after administration. During the chaos of the Cultural Revolution, researchers had to pursue this work secretly as universities and research institutes were closed, and faculty were often punished and sent to the countryside for so-called re-education.

Mao also charged practitioners of traditional Chinese medicine with searching ancient texts for possible medicines. An ancient practice using herbs and natural products, traditional Chinese medicine continues to this day in parallel to modern ("Western") medicine. Although rarely validated by modern science, traditional Chinese medicine includes thousands of folk remedies, of which a few may be useful. Mao's suggestion proved correct, and researchers located a 1,600-year-old manuscript composed by a physician, Ge Hong, that described the use of extracts from the sweet wormwood shrub (*Artemisia annua*) to treat recurrent fevers.

Quinoline, another anti-malarial, extracted from the bark of cinchona trees by Quechua Indians in Peru, has been used widely for generations to treat malaria. However, it has markedly unpleasant side effects, and access to cinchona bark is unreliable. First synthesized by German chemists, chloroquine proved more potent and better tolerated, with fewer toxic side effects than quinoline.

To reproduce the ancient method, an extract was first prepared and tested in mice and monkeys. Subsequently, the scientists used the extract to treat themselves and other humans and found rapid clearance of malaria parasites from the bloodstream. The active ingredient was isolated, and additional clinical testing confirmed remarkable efficacy.

Originally referred to in Mandarin as "qinghaosu," the compound is now known as "artemisinin." Artemisinin remained a secret medicine and was not published in the Chinese scientific literature until 1977, after the end of the war in Vietnam and the death of Mao Zedong. Artemisinin was not published in English until 1979. It was investigated by the World Health Organization in 1981, but it took years to persuade pharmaceutical companies to take an interest in a medicine for malaria, which was historically a disease associated with poverty. It then took until 2003 to get artemisin in approved to treat malaria in Zambia and subsequently throughout Africa.

Decades after initiating Project 523, the group leader, Tu Youyou, was recognized with the Lasker-Debakey Award for Clinical Medical Research (2011) and received the Nobel Prize for Physiology or Medicine (2015). Although she was a traditional Chinese medicine practitioner without training in modern medicinal chemistry and shunned scientists in modern laboratories during the Cultural Revolution, Dr. Tu followed through with the project Chairman Mao Zedong assigned her and changed the course of malaria worldwide.

FURTHER READING

The Lasker Foundation. "2011 Lasker-DeBakey Clinical Medical Research Award." 2011. https://laskerfoundation.org/winners/artemisinin-therapy-for-malaria/.

McNeil, Donald G., Jr. "For Intrigue, Malaria Drug Gets the Prize." *New York Times*, January 16, 2012. https://www.nytimes.com/2012/01/17/health/for-intrigue-malaria -drug-artemisinin-gets-the-prize.html.

The Nobel Prize. "Women Who Changed Science: Tu Youyou." https://www.nobelprize .org/womenwhochangedscience/stories/tu-youyou.

A self-described introvert, Thuma was emboldened by the study's results and became a champion for the new miracle medicine for malaria. He wanted every child, every patient, in Macha to receive the latest therapy. But there was a problem. The wonder drug would have to be used carefully. It has long been known that treatment with a single medicine gives infectious pathogens an upper hand because they quickly "learn" the toxic nature of the treatment and can shapeshift and evolve mutations to evade even the strongest therapies. This shape-shifting strategy can render newly discovered and once-promising drugs obsolete. What Thuma and his patients really needed was a cocktail of drugs that included the new miracle malaria medicine—barraging the malaria parasite with powerful attacks from various angles.

The absence of a cocktail of treatments was not the only problem. Because malaria is a disease of poverty, new malaria medicines were not deemed profitable by the pharmaceutical industry, meaning there was little chance of a drug company creating the combination Thuma sought. Eventually, one company, Novartis, developed artemether-lumefantrine (trade name Coartem),[28] which paired the quick-acting artemisinin compound with a second, slower-acting and long-lasting antimalarial drug. This combination came to be known as artemisinin combination therapy (ACT).[29] Given by mouth twice daily, the combination slammed malaria parasites with rapid and sustained agents.

Overjoyed that there was a new way to offer hope and relief to his patients, Thuma still faced a third, perhaps insurmountable impediment: The government. Armed with datasets, health records, and patient stories, Thuma repeatedly petitioned the Zambia National Malaria Elimination Centre (NMEC) staff to change the standard of care from chloroquine and quinine to ACT. The science was clear, but this was not an easy undertaking. Bureaucracies in government offices worldwide are often tradition-bound and concerned about economics, supply chains, and, in this case, a lack of familiarity with the pharmaceutical industry.

Refusing to accept a wait-and-see attitude, the usually shy Thuma presented his clinical data at meetings and continued to plead on behalf of his patients. Impressed with Thuma's data, the pharmaceutical company Novartis chose Zambia as its African partner nation for the drug's rollout. In 2003, a combination of medical diplomacy and lobbying by drug company executives made Zambia the first country in Africa to make the switch. ACT is now the standard treatment for malaria throughout sub-Saharan Africa.

Thuma did not anticipate the magnitude of this decision. The quiet life in rural Macha was disrupted by visits by dozens of news journalists and even pharmaceutical corporate executives who descended on the area. Shunning publicity had always been Thuma's response, but he was much sought-after. He did everything he could to concentrate on his work. Pressured to accept lucrative speaking invitations around the world and ghostwriters to expedite reports of the studies, Thuma told me that he had no interest in exposing himself and his family to unwanted attention.

The results were stunning. The Macha Mission Hospital's collaborative, multipronged efforts reduced malaria prevalence by 95 percent since 2004. No longer the leading killer in southern Zambia, childhood malaria—which had caused more than one thousand admissions to the children's ward of the Macha Mission Hospital, with up to a hundred deaths each year—plummeted to a few dozen admissions each year after 2004 when ACT was introduced.[30] The distribution of long-lasting insecticide-treated bed nets in 2006 further reduced the number of malaria cases.[31] Thuma and his colleagues went from watching scores of children die from malaria each rainy season to counting one or two deaths from malaria each year. Since 2018, no child malaria deaths have occurred at Macha Mission Hospital.[32]

The efforts were undertaken in coordination with the Zambia National Malaria Elimination Centre in the Southern Province. Key elements for the success at Macha were:

1. Screening for cases at the rural health clinics using rapid diagnostic tests and ACT treatment;
2. Follow-up visits to the patient's home for screening family members and neighbors; and
3. Distribution of long-lasting insecticide-treated bed nets.

Spraying of homes with insecticide was not utilized in the Southern Province.

I had met Chief Macha while staying in Zambia, and corresponded with him by emails that were printed and bicycle-delivered by Harry Hamapumbu. He had attended university to study accounting but returned to Macha as a farmer when he became chief. At age 78, Chief Macha has led his people for 31 years. While the chief was certainly pleased by the decline in malaria, I detected caution in his tone in past conversations and in his handwritten replies to my questions. He expressed gratitude to Thuma and his team but admitted concern for the future when Thuma would eventually retire and return to the United States. Malaria might return at that point, he said. "Malaria is a killer disease. It should be taught at secondary and university levels for future generations."

News of Macha's success brought massive hope to the region and beyond. If malaria could be fought in Macha, a rural town lacking significant infrastructure and miles from a paved road, then maybe all of Zambia could become malaria-free. The task, of course, was more complex than that. Malaria was not completely purged from Macha, and new cases would emerge among people who traveled outside the region. And still, despite the team's best boots-on-the-ground testing efforts, infections continued to spread from silent carriers of malaria parasites to small children who suffered from the debilitating disease.

News of Macha's 95 percent reduction in malaria cases sent ripples around the global health network. Malaria cases in other areas in central and southern Africa were also found to be declining, albeit less dramatically. Optimism that malaria was coming under control was voiced throughout Africa, but it was unclear what factors were responsible, and malaria did not decline everywhere. For example, the climate is warmer and very wet at Nchelenge in northern Zambia, on the border with the Democratic Republic of the Congo.[33] Mosquito populations there were resistant to standard insecticides; malaria transmission was enormous and did not decline after the introduction of ACT and long-lasting insecticide-treated bed nets.

From Baltimore to Botswana, government officials, economists, and scientists studied the Macha approach: A combination of tribal leadership, careful diplomacy, community engagement, advocacy efforts to multiple governments, and rigorous clinical research had emerged as the best solution to fight an ancient scourge. Unlike Soper, Thuma and his team did not employ sanctions and injunctions to test every person in the region. Instead, they relied on careful communication and respect for traditional customs. This, combined with the remarkable trust that the people in the region had for the Thuma family, who had lived in the community for decades, may have increased public awareness of malaria and willingness to accept medical care. Traditional healers using unscientific charms and potions still outnumber health care workers trained in modern medicines.

In 2005, an important funding source emerged—one of the most ambitious public health–malaria initiatives in history: the President's Malaria Initiative (PMI). With the goal of reducing malaria deaths by 50 percent across 15 high-burden countries in sub-Saharan Africa, PMI was initially launched by the administration of President George W. Bush with a $1.2 billion, five-year plan. Subsequent administrations renewed PMI; between 2001 and 2013, the initiative is estimated to have helped avert the deaths of 4.3 million people from malaria; more than 90 percent of these were children younger than 5 years.

MEDICAL DIPLOMACY OR SCIENTIFIC COLONIALISM?

US presidents have long focused on the health of foreign nations, even, and sometimes especially, after leaving office. But is this a millionaire's or billionaire's philanthropy, or a strategy for exerting control and interjecting continued interference on vulnerable populations under a benevolent guise? President Jimmy Carter established the Carter Center, which made ending Guinea worm its mission; President Bill Clinton founded the Clinton Health Access Initiative, focusing on infectious and noncommunicable diseases in the developing world; President George W. Bush launched PEPFAR, the President's Emergency Plan for AIDS Relief, and PMI, the President's Malaria Initiative, while in office. President Barack Obama launched the

Paul Farmer's New Model of Health Care Collaboration

When he was a high school student, Paul Farmer—the medical humanitarian whose legacy would become synonymous with a model of care that focused on social justice and equity—picked fruit alongside Haitian migrant workers. This work, and the experiences with those around him, sparked his lifelong interest in the Western Hemisphere's poorest country, and in the ways that medical care, or the lack of it, can perpetuate cycles of poverty.

Years later, while volunteering at a Haitian hospital and studying medical anthropology at Duke University, Farmer saw how the most impoverished patients couldn't afford access to life-saving medical care. The young anthropologist grew convinced that health care is a human right and that no one should be turned away from receiving what they needed.

Farmer grew up on an old school bus in a Florida trailer park that his father fitted with bunk beds. The second of six children, Farmer later lived on a leaky houseboat moored in an undeveloped bayou. His father, whom he described as a "free spirit," was a former teacher and frugal eccentric who always sympathized with the underdog—a trait that fostered an immunity to embarrassment in his children. His mother, a grocery store cashier, read the apartheid-era novel *Cry, the Beloved Country* to her children, who described her as "like the Virgin Mary without the virgin part . . . loving, kind, nonjudgmental."

While volunteering in Haiti, Farmer learned that he had been accepted at Harvard Medical School. He planned to read medicine and earn a doctorate in medical anthropology, but during three years of medical school, he commuted back and forth from Port-au-Prince to Cambridge, firm in his belief that the lessons learned while caring for impoverished patients were more valuable than the lectures delivered in the hallowed halls of the elite institution.

Harvard did prove useful in one regard; it was there that Farmer forged collaborations with classmates Jim Yong Kim and Ophelia Dahl. The trio launched Partners in Health in 1987, a medical nonprofit that approaches health care as a fundamental human right, and the establishment of robust health systems as social justice work. Farmer and his team believed in working with governments to train local workers and in building sustainable local clinics that could provide long-term care.

The Partners in Health model emphasizes medicine. It revolves around providing long-term, high-quality health care alongside comprehensive assistance in accessing food, transportation, and housing. Partners in Health is also driven by the idea that local capacity has to be built in order to establish continuity of care.

In its early days, the organization received financial support from Thomas J. White and Todd McCormack. In 1993, Farmer won a MacArthur "Genius" Fellowship for his vision. Working with government health ministries and local leaders, Partners in Health began with a single clinic in rural Haiti where locals were trained to become health care workers. The clinic grew over four decades to a staff of 18,000 people working with governments in 11 countries.

The Partners in Health model sees government collaboration as central to long-term impact and combines health care delivery with health advocacy and policy development. The organization collaborates with ministries of health in Rwanda, Lesotho, Peru, and Sierra Leone to develop sustainable care strategies, such as designing decentralized community-based mental health care systems. In Peru, Partners in Health has worked with the government to roll out safe houses for young women with severe mental illness who have been abandoned by their families. The Peruvian Ministry of Health later scaled the safe-house model nationally alongside another Partners in Health intervention, a chatbot for mental health screening.

In Rwanda, Partners in Health designed a health delivery model aligned with the government's mental health policy. Once the model was established and proven successful in one district, the Rwandan Ministry of Health scaled it to another district without the assistance of Partners in Health. The absence of Farmer's team was viewed as part of the mission's success, proving that Farmer's legacy was powerful beyond his presence.

Farmer's unexpected death in 2022, at age 62, sparked a renewed interest in his unlikely path from anthropologist to doctor to humanitarian, and served as a reminder that collaborations between physicians, scientists, and governments can lead to effective health care systems—the kind that no entity alone could successfully conjure or nurture.

FURTHER READING

Carmel, Julia. "Paul Farmer Is Awarded the $1 Million Berggruen Prize." *New York Times*, December 16, 2020. https://www.nytimes.com/2020/12/16/arts/paul-farmer -berggruen-prize.html.

Kidder, Tracy. *Mountains Beyond Mountains: The Quest of Dr. Paul Farmer, a Man Who Would Cure the World.* Random House, 2003.

Silver, Marc. "Dr. Paul Farmer Is 'Surprised and Upset and Humbled' After Visit to Haiti." *NPR,* October 21, 2016. https://www.npr.org/sections/goatsandsoda/2016/10/21 /498704601/paul-farmer-is-surprised-and-upset-and-humbled-after-visit-to-haiti.

Global Health Initiative in 2009 by funding $63 billion to consolidate US government programs.

The first executive director of the Global Health Initiative (2009–14), Lois Quam, was a former Rhodes Scholar with a unique career spanning private commerce (UnitedHealth), government service (Clinton White House), and faith-based aid programs (World Council of Churches). Recently, she was the CEO of the global nonprofit for reproductive health Pathfinder International.

Twelve years ago, reporting directly to US Secretary of State Hillary Clinton, Quam stressed that "the nation's focus on global health is a top priority for national security." Also, "Investing in the Centers for Disease Control is as important as investing in the Pentagon in terms of US national security. We are more likely to have lots of people die as a result of an infectious disease coming in than a military attack on US soil."

Looking back during our recent conversation, Quam shared additional perspectives. "It's so much more powerful if decisions can be made as close as possible to the work. So, we increasingly have key staff who work in countries outside of North America. That's a big and difficult shift, but we're making it. Trust is the coin of any realm. The work we do in reproductive health is very cultural."

"We really recognize women as agents . . . terrifically skilled people who have lots of thoughts about solutions," she said.

Critics of Western aid delivered under the pretext of medical diplomacy have described such efforts as scientific colonialism or medical imperialism, particularly when funding flows, repeatedly, to Western organizations tasked with fixing "Africa's malaria problem." In early 2021, when PMI launched a new, $30 million five-year research and evaluation plan, it was announced that the programs would be led by PATH. This US-based public health charity would work alongside seven other organizations, all centered in the United States, Australia, or the United Kingdom. This reignited old concerns about a lack of investment in local organizations run by African scientists and public health workers who better understand local dynamics.

Science has enabled Western colonialism for centuries, from advanced navigational instruments that allowed enslavers to chart oceans to quinine itself, which allowed colonials to endure the malaria-inflicted regions of Africa where they staked claim to the land of the native populations. Modern-day science includes so-called helicopter research,[34] also known as neo-colonial research or safari science, a phenomenon in which Western scientists travel to the Global South to extract data, publish the findings, and build scientific careers in a noncollaborative fashion for personal gain. One analysis found that 70 percent of randomly selected articles about lesser-developed countries published in peer-reviewed journals did not cite a local researcher.[35]

Such opportunistic behavior is counterproductive to the very goals of research and must be vigorously opposed. Turning again to the Macha experience, we are reminded that we have no right to enter another country, tell them what to do, and use their workers as subordinates. When Philip Thuma returned to Macha to build a malaria research program, he had already received approval from the US government and the Zambia National Malaria Elimination Centre, but he needed approval from the local government. To do so, he needed to meet repeatedly with Chief Macha and explain exactly what was planned and how this would benefit Chief Macha's people.

Moreover, Thuma exhibited commitment to furthering the careers of the young Zambian scientists and laboratory assistants. Every publication included the names of all the African team members. Taking the career ambitions of his younger colleagues seriously, Thuma advanced the careers of dozens of his young African protégés. Following his training at Macha, Godfrey Biemba went on to become director and CEO of the Zambian National Health Research Authority in Lusaka.[36] Modest Mulenga became director of the Tropical Diseases Research Centre in Ndola.[37] Sungano Mharakurwa was scientific director of MRT[38] and now is dean of the College of Health, Agriculture, and Natural Sciences at Africa University in Mutare, Zimbabwe.[39]

Lynn Paxton oversaw Tanzania's US-funded malaria-fighting efforts as the CDC's resident advisor for the US PMI from 2013 to 2017.[40] She said better efforts were needed to build local capacity when distributing millions of dollars of aid and that a centralized system ran the risk of misunderstanding the nuances within different communities.

"One of the wonderfully positive things about PMI was that we had figured out a formula of the basics," Paxton said.

"You had environmental interventions like spraying insecticides and mosquito nets; you had the treatment side, which meant getting drugs out to people; and then you had the diagnostic side. It had been figured out by someone in Washington that this is the core we want to deliver in all PMI countries. But each country has different needs." Paxton cites her own experience of navigating cultural and religious differences within Tanzania, where people living on the mainland are predominantly Christian,[41] while close to 99 percent of Zanzibaris are Muslim.[42] "On the plus side, it's helpful not having to reinvent the wheel for every country, but on the other hand, we had to figure out how to make this core work in Tanzania, which was like two countries in one and where there were very different social constraints on Zanzibar compared with the mainland."

Overall, US funding for malaria, which includes support for PMI as well as other malaria control and research activities, increased from $146 million in 2001 to $979 million in 2021.[43] PEPFAR, the President's Emergency Plan for AIDS Relief, experienced similar funding growth. Considered the brainchild of President George W. Bush and Anthony Fauci, PEPFAR has funded $100 billion in HIV/AIDS services since 2003, including support for the Global Fund to Fight AIDS, Tuberculosis and Malaria.[44] PEPFAR and PMI have saved an estimated 25 million lives in under-resourced nations,[45] highlighting the potential scope and impact of science diplomacy in action—as well as the cultural and sociopolitical challenges.

PEPFAR has been reauthorized three times since its inception,[46] each iteration raising new questions about the role and extent of US scientific interventions in developing countries. PEPFAR came under fire for its tightly focused, "vertical" approach to HIV/AIDS programming,[47] a strategy that specifies disease-specific goals. In this case, the focus was HIV/AIDS care, at the expense of other, perhaps more common or locally significant causes of death and suffering. And early iterations of PEPFAR demanded that recipients of US taxpayer funds spend some proportion of PEPFAR funds on sex abstinence programming,[48] a clause at odds with the science of HIV prevention and a cause of harm to those nonprofit and medical organizations caring for people engaged in sex work. PEPFAR officials also mandated that organizations sign anti-prostitution pledges,[49] excluding some of the most vulnerable women from receiving support and jeopardizing the existence of groups that worked with trafficked girls and women.

Paxton worked for decades on HIV/AIDS research in sub-Saharan Africa, which, while PEPFAR did not fund her, meant that she was "PEPFAR-adjacent" for a long stint. She avoided involvement with PEPFAR because the program was run in a paternalistic and bureaucratic fashion. "From the top down, it was mandated exactly how PEPFAR should be done, no matter what," she said. "But with PMI, things were better. PEPFAR was under Bush, whereas PMI came in a little bit later, and I think the imprint of whoever the first people are to start the program continues throughout. I wasn't there at the beginning of PMI, so I don't know exactly, but I get the sense that they had a different approach from the start."

The differences might stem from the stigma associated with the diseases themselves. "Malaria has never been as political as HIV. Ever. Not having that stigma attached to malaria where people in power say things like: 'You shouldn't shoot drugs, you shouldn't be gay,' that probably has always helped in terms of the palatability of the interventions proposed by PMI," said Paxton.

In 2010, Macha Mission Hospital received a share of the nearly $10 million in PEPFAR funds allocated to Zambia for HIV/AIDS research that year. Disseminated through a partnership with the Catholic Relief Services, these funds enabled Thuma's team to study how their patients took and tolerated HIV treatments. The team also studied whether home-based care helped people living with HIV/AIDS. One of the team's overarching goals, beyond ridding the region of malaria, was to prevent the typical helicopter-science approach and build local capacity by recruiting and retaining African scientists and leaders at Macha; this goal became a reality.

The Macha approach to reducing the burden of malaria disease has relied on careful medical diplomacy, respect for local norms and cultures, and meaningful community engagement and buy-in. It's a multipronged strategy that learns from historical figures like Fred Soper and William Gorgas and merges science and politics, yet works to avoid the kind of paternalistic, extractive science that can leave communities feeling exploited—a factor that can ultimately jeopardize longer-term partnerships. Macha's success might never have been realized without significant support from Western research aid. Yet support that comes with strings attached can, over time, perpetuate the very systems of oppression that force Global South countries into a cycle of reliance on outside experts.

SEARCHING FOR A VACCINE

There have been many attempts to develop a malaria vaccine, but until 2021 efforts were frustrated. Research at multiple institutions over 35 years, including the first pre-erythrocytic RTS vaccine (created in 1987), yielded the first successful vaccine, RTS,S.[50] The vaccine was tested at multiple sites in Africa and found safe for children. However, it required a series of four injections over several weeks, which was impractical for many patients, and the vaccine offered protection from severe disease for up to a year in only a third of vaccinated people.

A potentially improved recombinant vaccine, R21, was developed at the Jenner Institute at the University of Oxford and manufactured by the Serum Institute of India in 2023. This vaccine showed improved results in preliminary clinical trials. Twenty-eight African countries have requested the vaccines and plan to use them in tandem with their national malaria control programs. When used with long-lasting insecticide-treated bed nets, efficient screening, prompt treatment, and thorough follow-up, the vaccines are expected to save the lives of hundreds of thousands of young children in sub-Saharan Africa.

"As a malaria researcher, I used to dream of the day we would have a safe and effective vaccine against malaria. Now we have two," said Tedros Adhanom Ghebreyesus, director general of the World Health Organization. "Demand for the RTS,S vaccine far exceeds the supply, so this second vaccine is a vital additional tool to protect more children faster and bring us closer to our vision of a malaria-free future."

EMERGING THREATS

The fight to fund malaria control in Africa continues; 2023 brought an old threat back to the United States. Nine malaria cases were documented in the United States in 2023—the first such locally acquired infections in two decades. Seven people in Florida and one each in Texas and Maryland were among those who became infected, including a man who lived not far from Washington, DC, and left his home only once a day to walk his dog.

While malaria was endemic in the United States until the 1950s, it was eliminated in 1951. The malaria-spreading mosquito, however, has probably always been here. Our understanding of its habitats and behaviors within the US is limited because widespread surveillance is not routine.

The locally acquired infections likely occurred because travelers to South America and sub-Saharan Africa returned home to Florida, Texas, and Maryland with the malaria parasite inside their blood cells. The parasite was passed on to other Americans when local mosquitoes bit the malaria-infected travelers. The Maryland case was caused by *P. falciparum*. Cases in Texas and Florida were likely caused by *P. vivax*.

While *Anopheles* mosquitoes in sub-Saharan Africa are known to be more anthropophilic or human-loving (if you can call feasting on human blood "love"), *Anopheles* mosquitoes in the United States bite humans approximately 30 to 50 percent of the time. That percentage is closer to 98 in sub-Saharan Africa.

Still, news of Americans falling sick with malaria while never having ventured onto a plane might disrupt the 80:20 rule in public health. This principle describes how 80 percent of funding is spent on studying illness that afflicts 20 percent of the population, while the remaining 20 percent is invested in managing disease that affects 80 percent of the population. Long-standing inequities in malaria funding, which have enabled the disease's devastation in under-resourced countries for decades, may come to an end if Americans begin falling sick with malaria at home.

CHAPTER 5

Science on Trial

A COMMITTEE ON HUMAN RIGHTS

THE BATTLE OF SCIENCE VERSUS POLITICS and the machinations of the powerful who meddle with the work of scientists were on my mind in 2002 when I joined the Committee on Human Rights (CHR) of the US National Academies of Sciences, Engineering, and Medicine. At the time, I was a professor of biological chemistry at Johns Hopkins University. I felt moved to join the committee, knowing that my experiences working in and managing an American research lab were distant from the daily, dangerous, and existential threats facing some of my fellow scientists worldwide. I was yet to receive the Nobel Prize in Chemistry—that would happen the following year[1]—but I was already aware of the groundbreaking work of the CHR to protect scientists, engineers, and health professionals, much of it with the help of its members, including Nobel laureates.

The CHR is a staffed center that began its work in 1976,[2] around the time that Andrei Sakharov, the renowned Soviet nuclear physicist, was being exiled and admonished by Soviet officials because of his peace-building prodemocracy work. Sakharov was a global figure, well known for his pivotal role in developing the Soviet Union's first hydrogen bomb. But later, he became a human rights activist and a staunch advocate for the end of nuclear proliferation. The physicist was not afraid to speak out against injustice.[3] In 1964, he opposed the nomination of Nikolai Nuzhdin to the Soviet Academy of Sciences, of which Sakharov was a member, citing Nuzhdin's role in the "defamation, firing, arrest, even death, of many genuine scientists."[4]

Sakharov's vocal campaigning effectively blocked Nuzhdin's election. But as payback, the KGB began to collect *kompromat* on Sakharov. The situation only worsened three years later, on July 21, 1967,[5] when Sakharov petitioned the government, in a secret and detailed letter, to accept the US proposal for a bilateral rejection of antiballistic missiles.[6] The hope was that such an agreement would mitigate the possibility of global nuclear war, of which Sakharov was extremely fearful. Instead, the government banned Sakharov from conducting any research related to military activities, forbade him from speaking publicly about antiballistic missiles, denied his request to publish manuscripts on the topic, and, as he put it, "relieved me of my privileges in the Soviet Nomenklatura," the top strata of Soviet administrators.[7]

Sakharov's tireless and increasingly dangerous antinuclear and prodemocracy campaigning was noticed globally by concerned and appreciative scientists. When he won the Nobel Peace Prize in 1975, Soviet officials forbade him from traveling to Oslo to receive the accolade. Instead, Sakharov's wife, Yelena Bonner, traveled to Norway to deliver his acceptance speech.[8] Five years later, Soviet officials exiled Sakharov to Nizhny Novgorod in central Russia, a town notorious for the manufacture of nuclear submarines.[9]

American scientists watched the situation closely. Once Sakharov's internal exile began, members of the US National Academies of Sciences, Engineering, and Medicine organized letters of support and encouragement. Letters were sent to influential Soviet physicists, demanding that they publicly advocate for Sakharov's freedom of movement, freedom to conduct scientific experiments, and freedom to publish manuscripts.[10] Eventually, this effort grew to include members of the US National Academy of Engineering and the US National Academy of Medicine.

Since its inception, the CHR has contributed to the resolution of around 80 percent of the cases it has taken on, resulting in the freedom of more than a thousand scientists and health professionals.[11] This impact is due partly to the intellectual standing and political influence of the members of the National Academies. Scientists needing help are identified through professional organizations by committee members and staff and by direct appeal from their international colleagues. The CHR describes these workers as vulnerable to abuse and harassment "as a result of their evidence-based research, compliance with professional ethics requirements, and international collaborations, in cases where such activities are perceived as threatening by those in power. These professionals can also come under threat for refusing to use their expertise (e.g., by declining to participate in certain government projects). Many scientists and scholars are also subjected to severe ill-treatment as punishment for speaking out in support of justice and human dignity in their societies." [12]

The CHR analyzes each scientist's case to establish the authenticity and circumstances of the situation.[13] It verifies what charges the scientist faces, ascertains which laws were allegedly broken (through communication with family members, lawyers, and colleagues in the scientist's home country), and determines whether a fair (or unfair) trial has been held. Once validation is complete, committee members discuss the case and plan appropriate actions. In the past, this has included visiting scientists in prison, meeting with government officials, and attending the trials of accused scientists.

It has never been difficult to find scientists and academics who need CHR's support. But one case stands out for its complexity, the gravity of the charges, the international attention it drew, and the influential role of academic scientists in seeking justice for their incarcerated peers.

"CONFESS, OR YOU WILL DIE"

On January 29, 1999, 23 foreign medical staff members working at al-Fateh Children's Hospital in the eastern Libyan city

Anti-Science Policies

While the case of the Tripoli Six was ongoing, science came under attack by the US government, calling into question America's future as a scientific leader. In an effort to deter global terrorism after the 9/11 attacks, President George W. Bush quickly signed new laws that thwarted the very kinds of scientific partnerships we had promised to Libya in exchange for the health care workers' freedom. The ramifications of post-9/11 national security laws would incur long-standing damage to scientific progress.

While disciplines such as forensic research, cybersecurity, and biodefense flourished after injecting billions of dollars of funding, new and more rigorous regulations slowed progress, even in these areas. In the hopes of preventing pathogens from being used as infectious agents of war, for example, new import and export rules and harsher enforcement of existing policies posed labyrinthine challenges to researchers, discouraging some from continuing their studies in the life sciences.

The USA PATRIOT Act was especially detrimental to science. Signed six weeks after the 9/11 attacks, the PATRIOT Act enforced severe visa requirements for visitors to the United States that restricted scientists from attending conferences and continuing collaborations and halted the recruitment of science students by American universities and laboratories.

Albert Teich monitored the impact of those restrictions for the American Association for the Advancement of Science, where he served as director of science and policy programs. He said it can be difficult to precisely measure the direct impact of these visa changes on the ability of scientists to conduct experiments, forge collaborations, and present their work internationally, yet the net effect is negative and long term. Teich and his colleagues prepared two reports, including "Beyond 'Fortress America': National Security Controls on Science and Technology in a Globalized World," which was presented to Congress in the hope of creating a more conducive landscape for academic collaboration. The situation did slightly improve after those and other lobbying efforts, Teich said. The report outlined the impact of such legislation, stating that "the damage to US economic prosperity is significant." It also quoted a report of the Center for Strategic and International Studies' Commission on Scientific Communication and National Security: "In a world of globalized science and technology, security comes from windows, not walls."

But almost two decades later, Teich found himself confronting another barrier to US scientific progress. Teich was studying visas and immigration policies at George Washington University when President Donald Trump signed a January 2017 executive order banning citizens of seven majority-Muslim countries from entering the US. Thousands of doctors, scientists, and science students working and training in the US were left stranded when trying to return to the US after the winter break. Many more feared leaving the US in case they could not reenter. "It looks like we've gone back in time," Teich said a week after the Muslim ban was imposed. "This is an echo of what we did in 2002 and 2003. Back then, it was driven by panic. You could maybe understand it. What's happening now is a politically driven self-inflicted wound."

The Muslim ban had a particularly sharp impact on health care delivery in the US. More than a quarter of America's doctors hail from other countries, including around 8,400 who come from two countries listed in the executive order—Syria and Iran—according to the American Medical Association. The US faces a huge physician shortage, including a deficit of 8,200 primary care doctors and 2,800 psychiatrists, according to a 2016 report by the Association of American Medical Colleges. The shortage is expected to worsen, with the AAMC estimating a deficit of 94,700 physicians by 2025, almost a third of them primary care doctors. But under Trump's executive order, doctors from the listed Muslim-majority countries—many of whom received visas that required they practice medicine in rural and underserved parts of the US—were forced to leave. Not wishing to abandon careers in medicine, many resumed their training or clinical practice in the United Kingdom, Canada, and parts of the world that quickly recruited those no longer able to work in the US.

President Joe Biden revoked Trump's Muslim ban in 2021, allowing tens of thousands of doctors, scientists, and students entry into the United States. Many affected researchers and students already in the US on single-entry visas finally felt able to leave the country without fear of being blocked from returning. But in the space of four years, others had sought opportunities to advance science and medicine elsewhere, imposing a difficult-to-measure and likely long-standing negative impact on American science. Trump never apologized for the

Muslim ban or its detrimental impact on science; Bush did little to appease a scientific community that felt it was being punished because the masterminds of the September 11 attacks were graduates of the applied sciences. As of this writing—at the dawn of the second Trump administration—all signs point to a return to regressive immigration policies.

FURTHER READING

Mair, Michael. "NRC Issues Recommendations on Improving National Security Controls on Science and Technology." *Biosecurity and Bioterrorism* 7, no. 1 (2009): 1–16. https://doi.org/10.1089/bsp.2009.0213.

National Research Council. *Beyond "Fortress America": National Security Controls on Science and Technology in a Globalized World.* National Academies Press, 2009.

Sprenger, Sebastian. "Moscow's OK for Cooperative Security Liabilities Deal Expected Soon." *Inside the Pentagon* 21, no. 41 (2005): 1, 4–5. https://www.jstor.org/stable/insipent.21.41.01.

The White House, Briefing Room, Presidential Actions. "Proclamation on Ending Discriminatory Bans on Entry to the United States." Proclamation 10141, January 20, 2021. https://bidenwhitehouse.archives.gov/briefing-room/presidential-actions/2021/01/20/proclamation-ending-discriminatory-bans-on-entry-to-the-united-states/.

Yasmin, Seema. "Trump Immigration Ban Can Worsen US Doctor Shortage, Hurt Hospitals." *Scientific American,* February 1, 2017. https://www.scientificamerican.com/article/trump-immigration-ban-can-worsen-u-s-doctor-shortage-hurt-hospitals/.

of Benghazi were handcuffed, blindfolded, and beaten by police. The doctors and nurses, who hailed from Poland, Thailand, and other nations, were electrocuted, whipped with cables on the soles of their feet, attacked by police dogs, deprived of sleep, and suspended by their arms from interrogation room ceilings.[14] For two months, the health care workers were tortured and probed until most were released. But six of them—five Bulgarian nurses and an Egyptian-Palestinian physician—remained in solitary confinement. Some were confined as long as 10 months while their families struggled to locate the prisons in which their loved ones were jailed. For the next eight and a half years, the six health care workers were incarcerated in facilities in Tripoli and Benghazi. Their alleged crimes: deliberately infecting 426 hospitalized children with HIV. Their impending punishment: execution by firing squad.[15]

Libyan officials accused the foreign medical workers of secretly injecting HIV-infected blood from hidden vials into children as young as a few months of age in al-Fateh Children's Hospital in 1997 and 1998. The oldest "victim" was 14 years old. Many of the children had been hospitalized for breathing problems and other serious but non-life-threatening diseases, and most had received intravenous fluids and medicines; a few had been treated with blood transfusions.[16]

HIV rates in Benghazi were low, according to official figures, so the massive cluster of infections in one hospital was a medical mystery. Parents and Libyan officials, though, were adamant that the accused health care workers had deliberately injected children with HIV. They alleged that the doctor and nurses had taken payments from foreign scientists seeking to experiment on Libyan children in their quest to develop an HIV vaccine for use in the West. Some were spurred on by allegations made by Libya's leader, Colonel Muammar Gaddafi, that the medics were conspiring with the CIA and Israel's Mossad in an anti-Libyan plot.[17] The United States and Libya had been embroiled in a tumultuous relationship stretching back to the 1970s when the Gaddafi government funded anti-imperialist groups, including the Black Panthers in the US. The US imposed sanctions, and the relationship between the two nations quickly deteriorated. In the early 1980s, the US discovered Libyan-sponsored plans to assassinate American officials and attack American embassies. It launched airstrikes against Libya, Libya retaliated, and a hostile, violent relationship ensued.

The incarcerated health care workers were tortured into "confessing" that they had taken covert payments from foreign agencies. Over the course of months and then years, they were manipulated into saying that they had witnessed their peers knowingly inject children with HIV. The nurse accused of being the mastermind behind the entire scheme, Kristiana Valcheva, did not work in al-Fateh Children's Hospital at all; she worked in Al Hawari Hospital, having arrived in Libya eight years earlier.[18] But officials said bags of blood containing HIV were found in her home and that she had orchestrated the epidemic.

Another nurse, Valia Cherveniashka, said that despite the beatings and electrocutions, she had not confessed to committing any crime. It was Cherveniashka's husband who first brought the scandal to public attention when he staged a hunger strike outside the Libyan embassy in the Bulgarian capital in 2003. Alongside Cherveniashka and Valcheva was Valentina Siropulo, another Bulgarian nurse, who was beaten so badly that she was unable to speak for months and suffered partial paralysis of her face. In her first letter to her family after almost two years of imprisonment, Siropulo wrote, "Physically, I am relatively fine, but my soul is incurably ill." Another nurse, Snezhana Dimitrova, suffered from psychiatric illness and broken limbs.

Some of the nurses had arrived from Bulgaria only a few months before their arrest; most did not know one another. The Egyptian-Palestinian physician, Ashraf Ahmad Al-Hajouj, had begun his work as an intern at the hospital only five months before his arrest.[19] The torture inflicted on him during his incarceration left him blind in one eye and paralyzed in one hand.

Meanwhile, the 426 children who were infected with HIV were stigmatized in their communities and received inconsistent access to HIV medication and care. This wasn't a natural occurrence or a God-given disease, their parents argued; this was a deliberate act perpetrated against the children in a conspiracy to hurt the Libyan people. Many of the parents joined angry demonstrations outside the prison, chanting with protestors that the medics were guilty and deserved to die. Inside the prison, some of the health care workers continued to sign falsified documents as they were told to "confess or die." Cut off from the world, they were disoriented, confused, and scared for their lives. One nurse said that she was threatened with HIV infection and murder if she did not admit to spreading the virus.

The global HIV pandemic surged throughout the 1990s, orphaning children in sub-Saharan Africa and taxing already skeletal health care systems across the continent. Like other African nations, Libya looked outside its borders to recruit health care staff. As the third-largest oil exporter to the European Union, behind only Norway and Russia, the North African nation was wealthy enough to lure hundreds of nurses, doctors, and technicians each year from Eastern Europe, the Philippines, and other countries to care for Libyan patients. Salaries higher than those in their home countries attracted health care workers willing to relocate, including the Bulgarian nurses at al-Fateh Children's Hospital. Many said that the higher wages meant they could afford their living expenses in Benghazi and to send money to support family back home who relied on their help.

But as the HIV pandemic spread globally, health care settings became a significant source of viral transmission, particularly in low-income nations where equipment was limited and resources constrained. Single-use equipment was often reused on different patients, and scalpels and other medical implements were not properly sterilized. The virus was passed from one patient to the next by unsuspecting, overworked, or, in rare cases, negligent medical staff.

As the nurses and doctor awaited the death penalty in Libya, scientists from around the world, including the International AIDS Society, with its 12,000 members of academic AIDS specialists, caregivers, patients, and researchers, gathered evidence to explain how the outbreak might have occurred. Based on what was known about HIV clusters, especially epidemics that had occurred in hospitals in other parts of the world, international virologists and epidemiologists argued that the tragedy at al-Fateh Children's Hospital was a result of poor infection control and inadequate training of foreign medical staff. Health care workers might have the right nursing and medical training, they said. Yet those new to Libya were navigating a health care system and environment vastly different from those they had trained in, while tackling language and cultural barriers.

Supported by testimony from the virology world's most illustrious scientists, including French virologist Luc Montagnier, the co-discoverer of HIV, as well as international scientists, evidence was prepared to argue that the HIV outbreak in al-Fateh Children's Hospital was not the result of clandestine anti-Libyan activities. In fact, international experts said the outbreak had been noticed and reported *by* the health care workers themselves, and their concern triggered an in-person visit from World Health Organization officials in

1998. The public outcry to the revelation that 426 children had become infected in the hospital—news that spread to the Libyan public after the WHO visit—prompted government officials to frantically seek a scapegoat. They passed the blame for poor governance and inadequate resources onto the foreign nurses and doctor, continuing a pattern of blaming foreigners for domestic concerns.

The first hearing in the Benghazi HIV trial took place more than a year after the health care workers were arrested. The defendants filed into the Libyan People's Court, a special judiciary for cases deemed to impact national security. It was the first time that they had been formally charged with their crimes, which included deliberately injecting children with a deadly agent, attempting to destabilize the nation of Libya, and spreading a virus to cause an epidemic as part of an organized foreign conspiracy. They also faced charges of fornication, brewing alcohol, and violating foreign currency laws.

The nurses and doctor were granted access to a lawyer for the first time since their arrest. Those who had confessed to the charges retracted their confessions, noting that the confessions had been made while undergoing extreme torture.[20] This led to charges against eight Libyan prison guards and a Libyan physician and translator. Although medical examination of the defendants confirmed scars and evidence of electrocution and beatings, the guards and physician-translator charged with committing torture were acquitted. Some of the charges against the health care workers were also dropped. After multiple delays, the prosecution said it could not establish evidence for the attempted destabilization of Libya, so the People's Court declared itself unqualified to hear a trial that did not impact national security. This was not a victory; it would be another two years before the trial would resume after being transferred to Libya's Criminal Prosecution Service. Even then, it would be another year before the trial actually took place at the Benghazi Criminal Court in 2003. By then, 40 of the infected children had died from AIDS-related illnesses, and the health care workers had been incarcerated for nearly four years. The defendants again pleaded not guilty to deliberately infecting 426 children and retracted the confessions made while being tortured. They maintained their innocence for the duration of the trial.

All the while, Gaddafi was manipulating the Benghazi HIV trial for political gain. Having seized power in a military coup in 1969, Gaddafi, who hailed from western Libya, was seeking ways to increase his dwindling popularity in Benghazi and the Cyrenaica region of eastern Libya. At an international HIV/AIDS summit in Nigeria in 2001, he publicly accused a vast international conspiracy for intentionally infecting the 426 children in the al-Fateh Children's Hospital in an attempt to destabilize the country.

International scientists, including members of the Committee on Human Rights, rallied behind the health professionals, who became known as the "Tripoli Six." Between 2003 and 2006, arrangements were made for two distinguished scientists to travel to Libya to testify in support of the accused.[21] Vittorio Colizzi, a renowned Italian AIDS researcher from the University of Rome Tor Vergata, and Luc Montagnier from the Pasteur Institute in Paris (whose co-discovery of HIV would lead him to share the 2008 Nobel Prize in Physiology or Medicine) presented evidence to counter the prosecution's argument.

The process took three years because of court delays that required changing the scientists' schedules. Using rigorous genetic analyses, the pair showed that most of the children were infected with HIV during one year and concluded that the infections were due to poor hygiene and reusing improperly sterilized syringes and needles. By mapping the genetic origins of the viral variants found in the children's blood, the professors traced the infection to its likely source: a single unidentified child who was infected with HIV before April 1997. They also discovered that many of the 426 children were not only HIV-positive; they were co-infected with hepatitis B and C viruses.[22]

With strong genetic evidence and an epidemiological smoking gun, hope in the power of science to sway bureaucrats and convince a legal system was premature. Without explanation, the court disregarded the expert testimonies of Colizzi and Montagnier and ordered analyses by Libyan experts instead. The new "evidence" concluded that intentional infection of the children had taken place while the six health professionals were caring for patients at al-Fateh Children's Hospital.

More than a dozen foreign diplomats and international observers challenged the validity of the new evidence.[23] The data presented by Libyan experts was "inadequate and inconsistent," according to Janine Jagger, an epidemiologist who served as head of the International Health Care Worker Safety Center at the University of Virginia. Moreover, the Benghazi HIV trial failed to comply with international standards or even with Libyan law. Still, officials and many of the parents continued to support the notion of a surreptitious anti-Libyan plot. How else would 426 children become HIV-positive in the course of a few years?

Public pressure to indict the foreign health care workers was immense. Outside the court, local protestors continued to call for the execution of the nurses and doctor. The government of Bulgaria was offered the release of the six health professionals in exchange for a large cash settlement, but this was declined as it would imply the defendants' guilt. The parents of the HIV-infected children had each been offered a million dollars in compensation, but they declined because the money would come from the Libyan government, although they had been told that foreign conspirators were guilty of the whole episode.

Five years after the health care workers' initial incarceration, the Criminal Court reached its verdict. In May 2004, all six were found guilty and condemned to death by firing squad.

But the international scientific community was unwilling to concede. A year later, the case was brought before the Libyan Supreme Court, and an appeal was scheduled.

Gaddafi continued his exploitation of the HIV trial. Repeatedly, he singled out the CIA and Israel's Mossad as responsible for the pediatric HIV epidemic. The colonel made repeated comparisons to the perceived injustices of the trial in which a former Libyan intelligence operative was found guilty of planting a bomb on a Pan Am flight in 1988, killing 270 people over Lockerbie, Scotland. Abdelbaset al-Megrahi was sentenced to a minimum of 20 years in Barlinnie Prison in Scotland.[24] Gaddafi likened the Benghazi HIV trial to the Lockerbie trial, insisting that both were of international concern and that Libyan officials would impose justice on the nurses and doctor, just as international authorities had imposed so-called justice on al-Megrahi. Gaddafi's disinformation campaign only fueled Libya's public perception of the Tripoli Six. When the third Benghazi HIV trial began in August 2006, the prosecution again called for death by firing squad. Under Libyan law, retrials prohibited the release of new evidence, but in the seven and a half years since the health professionals were first arrested, worldwide interest had grown, as had evidence against the prosecution's arguments. International news media, in addition to the science media, followed the retrial closely and reported on the Tripoli Six. Jagger described the "molecular evidence" as backing the defendants and accused the prosecution of presenting a "shocking lack of evidence."

In December 2006, the new verdict was announced: guilty and condemned to execution. Again, outrage erupted outside Libya, particularly in Europe, where scientists and scientific organizations released scathing criticism of Libya and the Gaddafi regime. Yet the public outrage seemed to have no discernable effect on the Libyan government. Robert Gallo, a virologist and pioneer in HIV research at the University of Maryland, organized a letter demanding justice for the accused, which was signed by 43 of the world's leading HIV experts. Renowned scientific organizations, such as the British Royal Society and the Royal College of Physicians, published pleas in the *London Times*.[25] But these public attempts at scientific diplomacy, along with hundreds of letters sent to Gaddafi by members and co-chairs of the CHR, were met with silence. We wondered whether Gaddafi had even learned of our appeal. If public pleas were futile, a new strategy was needed: private meetings between Gaddafi, Gaddafi's son, and two eminent scientists. I composed a letter, delivered to Colonel Gaddafi by hand:

February 8, 2006

His Excellency Colonel Muammar al-Gaddafi
Leader of the Revolution
Office of the Leader of the Revolution
Tripoli, Socialist People's Libyan Arab Jamahiriya

Excellency:

I am writing in my capacity as chairman of the Committee on Human Rights of the US National Academy of Sciences, National Academy of Engineering, and Institute of Medicine. We welcome the December 25, 2005, decision of Libya's Supreme Court to overturn, on appeal, the death sentences that were pending against five Bulgarian nurses (Kristiana Malinova Valcheva, Nasya Stojcheva Nenova, Valentina Manolova Siropulo, Valya Georgieva Chervenyashka, and Snezhanka Ivanova Dimitrova) and a Palestinian medical doctor (Ashraf Ahmad Jum'a [Al-Hajouj]) and to order a retrial. It is our understanding that the retrial will be held shortly before the Benghazi Criminal Court.

The HIV infection of hundreds of children and the deaths of about 50 of them in Benghazi is a tragedy. The plight of those who are infected merits significant medical and humanitarian assistance. To that end, we are gratified by reports that, through the combined efforts of Libya, Bulgaria, the European Union, and the United States, the International Benghazi Families Support Fund has been created to provide continuing medical assistance to the infected children, to upgrade the clinic in Benghazi where the children receive medical treatment, and to provide financial support to the children's families.

We are concerned, however, that the Bulgarian nurses and Palestinian medical doctor named above have already been incarcerated for more than seven years despite scientific evidence, provided at their trial by internationally renowned scientists Luc Montagnier and Vittorio Colizzi, that reliably indicated that the infections were the result of poor hospital hygiene practices. Furthermore, we understand that scientific documentation was also provided showing that some of the HIV infections predated the nurses' and doctor's arrival at the Benghazi hospital and others occurred after they were arrested.

Given the above, we are confident that, if the retrial is held before an independent, open, and impartial tribunal—in accordance with international fair trial standards—the defendants will be released from prison. Because of the defendants' already lengthy incarceration, we respectfully request that, for humanitarian reasons, the trial be thorough but expeditiously conducted.

Thank you for your attention to this matter of importance to us and to many other members of the international scientific community.

Sincerely,

Peter Agre, M.D.

Chairman

(2003 Nobel Laureate in Chemistry)

cc: Mr. Terry Davis, Secretary General, Council of
Europe
The Honorable Condoleezza Rice, Secretary of State,
US Department of State
(Letter also sent to Minister of Justice Ali Hasnawi.)

Robert Putnam, the political scientist and former dean of Harvard University's Kennedy School of Government, would meet with Gaddafi. Richard Roberts, the British Nobel laureate, biochemist, and chief scientific officer at New England Biolabs, would meet with Gaddafi's son, Saif al-Islam.

In early 2007, Putnam, a member of the National Academy of Sciences and an advisor to several US presidents, received an invitation from the Libyan government to visit top leaders in Tripoli to discuss the future of Libyan civil society. He was invited to discuss his research on civil society and his book, *Making Democracy Work.* Putnam contacted Paula Dobriansky, US Under Secretary of State for Democracy and Global Affairs, and William Colglazier, Executive Officer of the National Academy of Sciences, who had both been frustrated by their inability to meet with Gaddafi during a 2006 trip to Libya. Dobriansky and Colglazier strongly encouraged Putnam to accept the invitation and to use the opportunity to appeal the case of the Tripoli Six.[26]

Colglazier shared his reflections on Libya and the potential of science diplomacy. He told me,

When Gaddafi agreed to give up his program for nuclear weapons, the [US] State Department was trying to re-engage with Libya to see what constructively could be done, in terms of providing link-ups with academic scientists at the universities and Libya, to use science technology to try to help improve the lives of its own people.

I had been before in Tripoli and Benghazi where we had a number of meetings with [Libyan] academics. What I remember most prominently of the dinners was the academics who came up privately to say that the best time in their life was when they were getting their PhD at Kansas State or at the University of North Dakota. I think the US did not realize the full potential that it could have using science as a very positive diplomatic tool for engaging with countries.

A meeting with Gaddafi was the only chance to hand-deliver letters and discuss the imminent execution of the six health professionals face-to-face. It was decided that Putnam would present a letter to Gaddafi from the presidents of the National Academies. The letter conveyed the importance of scientific collaboration in Libya's future growth but said the release of the six health care workers would be essential to fruitful international scientific partnerships. The presidents wrote of their interest "in expanding collaborations and joint activities between US and Libyan scientists, engineers, and medical professionals," promising that science could "help to solve important problems facing society and improve the lives of our citizens." The letter carefully expressed sympathy for the Libyan children infected with HIV but closed by stating that scientific evidence did not support the argument that the children had been intentionally infected.

Putnam's trip to Libya was one of the most unusual missions the CHR had ever assisted. By the time of Putnam's visit, Gaddafi had clearly signaled that he wanted to resume normal relations with the European Union and the United States but argued that the incarceration of a Libyan citizen for the Lockerbie bombing would not be forgotten.

Putnam flew to Libya with his wife, Rosemary, and toured historic sites. The couple was invited to a private dinner at the home of Abdullah al-Senussi, head of the Libyan Defense Intelligence, a longtime confidant of Gaddafi, and Gaddafi's brother-in-law.[27] Having been found guilty in absentia for his role in the Lockerbie bombing, al-Senussi was notorious abroad but highly influential within Libya. Putnam showed al-Senussi the letter from the presidents of the three National Academies, and they discussed Libya's future and international standing.

The next day, the Putnams were taken on a private jet to Sirte, the provincial town on the coast close to Gaddafi's desert residence. Flying with him were Libyan Prime Minister al-Baghdadi al-Mahmoudi and Omran Bukhres, a Libyan-American who described himself as a mentor to Gaddafi's son, Saif al-Islam. The Putnams waited all day, but no meeting took place. Instead, apologies were issued without any explanation of Gaddafi's absence. But while he waited, Putnam was asked by one of Saif al-Islam's aides whether he would edit Saif al-Islam's thesis on economics, which was being prepared for his PhD at the London School of Economics. A puzzled Putnam looked over the thesis but said he couldn't edit the document because the subject was outside his field of expertise. Putnam held back his true thoughts: The writing was academically fraudulent, and the thesis appeared to have been ghostwritten. Eventually, that same thesis would result in the dismissal of senior leadership at LSE over accusations of plagiarism.[28] Putnam never did meet Saif al-Islam, but that exchange underscored his impression that Gaddafi's son was an opportunist who used oil wealth to burnish his Western credentials.

After a frustrating day with no Gaddafi in sight, the Putnams flew back to Tripoli. The next morning, the Putnams were whisked directly from Sirte Airport at high speed to a compound framed by a large wall enclosing about one square mile of desert. Inside was Gaddafi's large Bedouin tent surrounded by a field of wildflowers, lavender, and eucalyptus trees with a small herd of grazing camels. The compound gave the illusion of a pastoral scene, but parked nearby were telecommunications vans and patrolling bodyguards with guns.

Gaddafi was listless and disheveled. The 64-year-old revolutionary had run Libya for 37 years, and Putnam found himself drawing comparisons between the colonel and a fatigued Mick Jagger. But gradually, Gaddafi perked up and became attentive and intellectually coherent. The meeting exceeded the two hours planned, and their discussions focused heavily on democratic reform in Libya. Gaddafi steered the conversation to his 1975 publication, *The Green Book,* which described his political philosophy. Putnam considered it "an archaic mixture of primitive socialism, 1960s-style 'people power' rhetoric, and traditional Bedouin values; it has been the touchstone and straitjacket for politics in Libya for nearly four decades." It had become clear that Gaddafi considered himself a profound thinker. With a straight face he referred to himself and Putnam as "philosopher-kings."

"I smiled," Putnam recalls, "and was at a loss for words. Colonel Gaddafi was a tyrant and a megalomaniac, not a philosopher-king. But our visit left me convinced that he was not a simple man." Hours into the conversation, Putnam reached into his pocket and took out the letter from the three presidents of the National Academies as well as its Arabic translation. Gaddafi became very quiet as he pored over promises about scientific collaborations and the gleaming possibility of a desalination facility alongside other modern infrastructure projects in Libya. The letter from the presidents was clear: Any plans for partnership were dependent on the reprieve of the death sentences for the five Bulgarian nurses and the Egyptian-Palestinian physician. Eventually, Gaddafi admitted to a desire for international collaboration. He never argued about the foreign health care workers' guilt, innocence, or intentions. Instead, he said, "This is a good idea. We will look at it closely. Collaboration is a good idea, and we will take [the letter] seriously."

Then, in a complete surprise, Gaddafi launched into a harangue about what he considered the biggest problem facing the Middle East: Saudi Arabia and its dominant form of Islam, Wahhabism, a conservative and influential branch within Sunni Islam. Gaddafi delivered his rant with such force that Putnam wondered whether this was the real reason for the meeting. Sensing that Gaddafi was still hostile to US oil interests and US politicians, he chose not to probe further. Gaddafi said that Putnam should deliver this important message to "your friends in the State Department." The two signed books, posed for photos, and said farewell. The meeting felt like a success.

The Putnams spent the weekend visiting Ghadamès, a UNESCO World Heritage oasis and ancient city in the desert, and returned to Tripoli on January 21, where they had a final meeting with al-Senussi at his office in the Libyan Defense Intelligence Agency. The Putnams and al-Senussi compared their favorable impressions of the meeting with Gaddafi two days earlier outside Sirte (which had already been reported to al-Senussi). Having eliminated weapons of mass destruction, Libya expected help to develop the country. But the letter from the presidents of the National Academies and the discussion between Putnam and Gaddafi made it clear that collaboration on projects such as the desalination plant could not go forward until the issue of the Tripoli Six was resolved. Al-Senussi directly leveled with Putnam that the government of Libya needed help, as the families of the HIV-infected children had refused to accept money from their own government, believing the endless propaganda that the perpetrators of this great crime were the foreign health professionals. Al-Senussi requested help from philanthropists or nongovernmental agencies and the National Academies. Putnam reminded al-Senussi that he had no authority to negotiate financial transfers.

Putnam found al-Senussi smart, shrewd, and focused on getting things done. Al-Senussi was very close to Gaddafi's son, Saif al-Islam, and seemed to strongly support Libyan reform. Perhaps revealing a viewpoint similar to that of Saif al-Islam, al-Senussi said that he believed in education and asked for details about admission to the program for policy studies at Harvard's Kennedy School, where Putnam was dean; Al-Senussi hoped his son would be admitted. Putnam left Libya after a surreal week, uncertain about the future of the health care workers but hoping for their peaceful release.

Four years later (2011), in a commentary in the *Wall Street Journal*, Putnam reflected on the meeting with Gaddafi: "Was this a serious conversation or an elaborate farce? Naturally, I came away thinking—hoping—that I had managed to sway Col. Gaddafi in some small way, but my wife was skeptical. Two months later, I was invited back to a public roundtable in Libya, but by then, I had concluded that the whole exercise was a public-relations stunt, and I declined."[29]

The second scientist to privately meet with the Gaddafis was the British Nobel laureate and longtime chief scientific officer of New England Biolabs, Richard Roberts.[30] A series of 2006 news stories about the Benghazi HIV trials and the imminent execution of the health care workers had caught Roberts's attention.

Roberts recognized that it was very unlikely that patients in al-Fateh Children's Hospital had been deliberately injected with HIV, and he penned an open letter to Gaddafi, which was signed by 113 fellow Nobel laureates and published online in the journal *Nature* in 2006. The letter raised objections to the Libyan court's dismissal of evidence from expert scientists that would have exonerated the Tripoli Six.[31]

Roberts had been practicing science diplomacy since long before the plight of the Tripoli Six. During the Cold War, he visited Russia to meet with fellow scientists and began to understand the power of conversations between international scientists. "In Russia, I would meet mainly with scientists, but occasionally with politicians. And I realized that the only way that scientists in Russia were getting news about what was going on outside of their country was from visiting scientists." He also made annual trips to Pakistan for science meetings to which government officials were invited. "It was really to convince the government to invest in science and to recruit Pakistani [students] to work in labs in the US, UK, and elsewhere," he said.

Since the letter garnered no response from Libyan officials, Roberts went to New York in October 2006 and knocked on the door of the Libyan delegation to the UN General Assembly. A seasoned debater, Roberts eventually found his way to a meeting with Libya's ambassador to the United Nations, Attia Mubarak, to whom he presented the letter signed by the Nobel laureates.[32] The ambassador listened politely for an hour and a half as Roberts stated his objections. But Mubarak voiced regret that Roberts had already gone to the press with his concerns. This approach went against the tradition of Islamic law and diplomacy, he said.

Deeply frustrated, Roberts contacted the US State Department for assistance in escalating his concerns, but the conversation met a dead end when State Department staff said they couldn't help because of political tensions between the United States and Libya. Roberts contacted the UK Foreign Office in London. As Roberts was a British Nobel laureate, a member of the Royal Society, and on the Queen's honors list to receive a knighthood, the Foreign Office took him seriously and promised to help. A few weeks later, he received a call from the chief of staff for Saif al-Islam, Gaddafi's son, who said he had read his letter and could sympathize. "It was unexpected," said Roberts. "The call came out of the blue, and a few days later, I was on a plane to Tripoli."

The British Foreign Office arranged for Roberts to stay at the British Embassy in Tripoli. Roberts arrived and waited for a call. "At 10 P.M. I received a call saying a car would take me to meet Saif al-Islam." The pair talked alone for 30 minutes in a hotel room. Saif al-Islam had read the letter and asked Roberts to elaborate on how the hospitalized children might have unintentionally been infected with HIV. "I told him that because needles were shared in Libyan hospitals, this was the likely reason for the outbreak. He listened carefully." To Roberts's surprise, the discussion ended with Saif al-Islam making a promise regarding the Tripoli Six: "We know that what you're saying is true, and we're going to let them go." Roberts was shocked. "Saif al-Islam said they would be released quickly."

Roberts felt fantastic—until he returned to the United States, where a letter from the UK Foreign Office described a "slight wrinkle" in the situation. The diplomatic situation was becoming increasingly complicated: the newly elected French president, Nicolas Sarkozy, had made campaign promises to rescue the imprisoned health care workers, and his wife, Cecilia, wanted to fly to Tripoli to accompany them back to Bulgaria. But on hearing this, Libyan officials demanded a large sum of cash, and Roberts felt this was out of line given the promise made to him by Saif al-Islam. "It was a political stunt," said Roberts. "The Brits were really pissed off. I think what happened is that when some Libyan officials heard that other nations were getting involved, they decided to try and get some cash out of it."

Roberts reached out to Saif al-Islam to discuss the matter. Saif al-Islam then asked Roberts to serve as a scientific advisor on his charitable foundation's board. Roberts agreed. But the promising midnight conversation in Tripoli and his attendance at the foundation's meetings didn't seem to help. The Tripoli Six remained stuck.

RETRIAL AND RESOLUTION

The retrial began in June 2007. The Supreme Court of Libya began deliberations as tensions rose, and there was massive international disappointment when the verdict was delivered in July. Twenty families of infected children protested outside of the court where the Tripoli Six were still deemed guilty and still condemned to execution. But then the unexpected happened. Saif al-Islam's foundation, with its self-declared mission of supporting human rights, reached a settlement with the families of the infected children and with the European Union, which agreed to provide therapeutic treatments and $1 million to the families of each child.[33] Under Islamic law, the families were allowed to grant clemency in exchange for compensation.

The Libyan Supreme Judicial Council, Libya's highest legal body, commuted the death sentences to life in prison. Bulgaria had joined the European Union on January 1, 2007, and

President Georgi Parvanov granted honorary Bulgarian citizenship to the Egyptian-Palestinian physician, Ashraf Ahmad Al-hajouj.[34] A bilateral agreement between Libya and Bulgaria allowed citizens of one country convicted of crimes in another country to serve their sentences in their home nation. Benita Ferrero-Waldner, the European Commissioner for External Relations, flew to Tripoli with Cecilia Sarkozy in the French president's plane to accompany Al-Hajouj, Kristiana Valcheva, Nasya Nenova, Valya Chervenyashka, Valentina Siropulo, and Snezhana Dimitrova on the flight to Sofia.[35] When they landed on Bulgarian soil on July 24, 2007, President Georgi Sedefchov Parvanov pardoned the Tripoli Six. Putnam, Roberts, and others involved in the long advocacy campaign were relieved, even if they were unsure which element of the yearslong battle was responsible for the eventual bilateral agreement. "I was pleased," said Putnam. "But I had then and have now no idea whether my negotiations with Muammar Gaddafi and his colleagues played any actual role in their release and pardon."

Back at home on the East Coast of the United States, Roberts, whose midnight conversation with Gaddafi's son had been full of promise, celebrated with scientists. "We had a few bottles of wine to celebrate because when I first went to Libya, I bought my own ticket, but my company had offered to pay for it, so my colleagues were well aware of the situation. They felt invested in the situation as well."

But Roberts didn't claim victory for the scientists-turned-advocates. "The reason it all worked out is not because we scientists were forcing it. It's because we were able to talk to diplomats, in my case, the Foreign Office in London, and convince *them* to do the right thing. It wasn't on their radar. We were able to add legitimacy to the things that diplomats can do—especially as Nobel laureates."

The case of the condemned health professionals was finally over, but only after prolonged imprisonment, torture, agony, false allegations, denied justice, and far-reaching condemnation. The nurses and doctor suffered eight and a half years of incarceration and physical and psychological trauma, while their families endured nearly a decade of anguish. Of the 426 children, more than 62 had died by 2011. At least 19 babies born to those infected in al-Fateh Children's Hospital became HIV-positive through breastfeeding.

Decades after the outbreak, parents of the affected children began to voice, for the first time, a different theory from their original claims. They said that perhaps the Gaddafi regime had manipulated the epidemic and put the blame on foreign medical workers as a form of retaliation against the incarceration of a Libyan national for the Lockerbie bombing. Some parents referred to the millions of dollars they received in compensation as blood money, unsure of the cash flow's provenance. There was talk that it came from the European Union, but there was a deep distrust around government bureaucracies and a perception that Saif al-Islam had laundered his own funds through other agencies.

In 2011, Libya descended into civil war between forces loyal to Gaddafi and opposing rebel groups. Gaddafi was in a convoy west of Sirte that came under NATO attack. He was captured and killed by National Transitional Council forces on October 20, 2011. Saif al-Islam was captured and imprisoned in Zintan in November. Calls for his extradition to the International Criminal Court in The Hague, to face charges of war crimes, were ignored. Saif al-Islam remained in confinement, resurfacing in 2021 when he announced that he would run for president.[36] In March 2021, former French President Nicolas Sarkozy, whose wife had accompanied the freed health care workers, was convicted of corruption and influence peddling. In September, he was convicted of illegal campaign financing during his unsuccessful 2012 presidential campaign.[37] He was sentenced to one year in jail.

Scientists played a vital role in the eventual release and pardon of the Tripoli Six, even though Roberts and others questioned the exact and tangible nature of their influence. The collaboration of scientists with diplomats, who were mostly unaware of the health care workers' detention and torture, is often credited as pivotal, and Roberts said he uses the campaign to highlight to younger scientists how science diplomacy can function. "I do try to convince people that science diplomacy is worthwhile," he said. "Nowadays, scientists try to avoid anything they consider political. But what you can do as a scientist is talk truth to power. Diplomats can't do that. Diplomats have to toe the party line of the country they're coming from, so they can't be honest as scientists can. Scientists have a worldwide network, and we have shown that we can solve problems without guns."

Epilogue

THE WEALTHIEST AND MOST POWERFUL nation on Earth suffers from an image problem when viewed from abroad. Negative perceptions of the United States abound and have been aggravated by the statements and policies of some US politicians and their minions: purging federal health agencies, shuttering international health aid, referring to "shithole" countries in Africa, banning the entry of Muslim scientists and health professionals into the United States, and recklessly tightening sanctions on other nations are just a few examples. In 2004, Zogby polls of five Arab nations indicated that the United States is viewed as arrogant, untrustworthy, and easily provoked, with fewer than a fifth of respondents finding anything positive to say.[1] In contrast, these same respondents expressed a 48 to 90 percent approval of US science and technology—a pathway with potential for improving relations and perceptions, although one currently under a real and also existential threat in the US given staffing and grant fund cuts.

Science is a highly respected human endeavor. While it sometimes tumbles into the crosshairs of sanctions and trade restrictions between warring countries, it fosters international travel and collaboration all the same. Despite recent challenges raised by US rhetoric and policies, young scientists from around the world pipette, centrifuge, and peer into microscopes in laboratories across America. They dream of careers filled with scientific discoveries that improve the health and well-being of people and the planet. Science is also a life-changing endeavor, one that fosters enduring international friendships, like the story of Ernest Moniz and Ali Akbar Salehi, physicists who trained at MIT in the 1970s and later played pivotal roles in negotiating the 2015 Iran Nuclear Accord.

In my case, winning a Nobel Prize in 2003 opened the door to invitations to visit universities around the globe, including countries that have been unwelcoming to the United States for decades: Cuba, Iran, North Korea, Libya, and nations in sub-Saharan Africa (Zambia, Zimbabwe, and the Democratic Republic of the Congo). The award led to meetings with heads of state, including France's President Nicolas Sarkozy, Iran's President Mahmoud Ahmadinejad, Cuba's Presidents Fidel Castro and Miguel Díaz-Canel, Japan's Prime Minister Shinzo Abe, and India's Prime Minister Narendra Modi. While these leaders held dramatically different political biases and viewpoints, all were convinced that scientific endeavors were critical for both the survival and the prosperity of their nations.

I hope that respect for science and its role in human advancement will build bridges between countries, reduce animosities, and lead to international cooperation on universal problems such as global climate change, mass extinctions, air and groundwater pollution, emerging pandemics, and diseases, including Alzheimer's and cancer, even as the United States has recently taken steps to disengage from international cooperation. Science will never eliminate the need for a diplomatic corps, but the roles of science and scientists are of significant value to America's foreign policy and warrant greater emphasis. In particular, international exchanges of young scientists will be important, for among them are the future leaders of global science—and the future brokers of peace.

Acknowledgments

PETER AGRE

I am deeply indebted to the American Association for the Advancement of Science for introducing me to Norman Neureiter and Vaughan Turekian, who founded the Center for Science Diplomacy and stoked my passion by including me on multiple international adventures.

My involvement in human rights was catalyzed by the Committee on Human Rights of the National Academies, led at the time by Torsten Wiesel and Carol Corillon, and by Bill Colglazier.

I thank Michael Bloomberg and Bloomberg Philanthropies for funding the Bloomberg Distinguished Professorships and the Johns Hopkins Malaria Research Institute (JHMRI) where I served as director for 16 years and oversaw our field work in Zambia, Zimbabwe, and the Democratic Republic of the Congo (DRC). Additional support was provided by Maxmillian Angerholzer III of the Richard Lounsbery Foundation, who traveled with us to Cuba and the Democratic People's Republic of Korea.

Key individuals facilitated entry and fostered working relationships in each of the countries visited. In Cuba we were hosted by the late Fidel Castro Díaz-Balart, Jorge Pastrana (Cuban Academy of Sciences), and Luis Alberto Montero Cabrera (University of Havana). The Islamic Republic of Iran visit was arranged by Ali Akbar Salehi (nuclear physicist and former foreign minister), Hassan Vafai (University of Arizona), and the late Glenn Schweitzer (National Academy of Sciences). James C. K. Kim and Chan-Mo Park (Pyongyang University of Science and Technology) arranged visits to North Korea. Philip Thuma (Macha Research Trust), Sungano Mharakurwa (Dean, Africa University), and William J. Moss (Johns Hopkins) arranged our efforts in Zambia, Zimbabwe, and the DRC.

This book would not be possible without support from Johns Hopkins's Vice Provost of Research, Denis Wirtz; his team in the Office of Research; Executive Director Barbara Kline Pope and her team at Johns Hopkins University Press; the encouragement and direct editorial involvement of Hopkins's Director of Strategic Engagement, Anna Marlis Burgard, and my coauthor, Seema Yasmin of Stanford University.

I am grateful to my parents for encouraging me long ago to follow my dreams even though it meant permanently leaving my home state, Minnesota. Most of all, I am entirely beholden to Mary, my wife and life partner. In addition to being the mother of our four wonderful children and grandmother of our six irresistible grandchildren, she has kept me on track since we first met and fell in love 50 years ago.

SEEMA YASMIN

Big thanks to Lilly Ghahremani and Anna Marlis Burgard for their careful shepherding of this project.

Notes

PREFACE

1. Council on Foreign Relations, "U.S.–Cuba Relations, 1959–2023," accessed December 4, 2024, https://www.cfr.org/timeline/us-cuba-relations.

2. Paul K. Drain and Michele Barry, "Fifty Years of U.S. Embargo: Cuba's Health Outcomes and Lessons," *Science* 328, no. 5978 (2010): 572–73, https://doi.org/10.1126/science.1189680.

3. The White House, "Charting a New Course on Cuba," accessed March 12, 2025, https://obamawhitehouse.archives.gov/issues/foreign-policy/cuba.

4. Sarah Zhang, "Cuba's Innovative Cancer Vaccine Is Finally Coming to America," *The Atlantic,* November 7, 2016; John M. Kirk, "Cuba's Medical Internationalism: Development and Rationale," *Bulletin of Latin American Research* 28, no. 4 (2009): 497–511, https://doi.org/10.1111/j.1470-9856.2009.00314.x.

5. US Senate, "Sputnik Spurs Passage of the National Defense Education Act," accessed December 4, 2024, https://www.senate.gov/artandhistory/history/minute/Sputnik_Spurs_Passage_of_National_Defense_Education_Act.htm.

6. Douglas O. Linder, "The Security Hearing of Robert Oppenheimer (1954): An Account," *Famous Trials,* University of Missouri–Kansas City School of Law, 2020, https://famous-trials.com/oppenheimer/2688-the-security-hearing-of-robert-oppenheimer-an-account.

7. "Science and Foreign Relations: Berkner Report to the U.S. Department of State," *Bulletin of the Atomic Scientists* 6, no. 10 (1950): 293–98, https://doi.org/10.1080/00963402.1950.11461292.

8. "Benjamin Franklin Sets Sail to France," *History.com,* accessed February 1, 2023, https://www.history.com/this-day-in-history/benjamin-franklin-sets-sail-for-france.

9. University of California, Berkeley, College of Chemistry, "In Memoriam: Alum Harvey Itano and the Journey to Sickle Cell Research," accessed December 4, 2024, https://chemistry.berkeley.edu/news/memoriam-alum -harvey-itano-and-journey-sickle-cell-research.

CHAPTER 1. CUBA

In addition to the sources listed below, we have benefited from conversations with Rush Holt Jr., former chief executive officer of the American Association for the Advancement of Science and former eight-term US Representative (D-NJ); Luis Alberto Montero Cabrera, former president of the Cuban Chemical Society and professor of chemistry, Universidad de La Habana; Sergio Jorge Pastrana, former executive director and secretary of foreign affairs of the Academia de Ciencias de Cuba; Manuel Raíces Pérez-Castañeda, Cuban Center for Genetic Engineering and Biotechnology; Mark Rasenick, Distinguished Professor of Physiology & Biophysics and Psychiatry, University of Illinois at Chicago School of Medicine; and Pedro Antonio Valdes-Sosa, vice-director for research at the Cuban Neurosciences Center.

1. Manuel Amador and Manuel Peña, "Nutrition and Health Issues in Cuba: Strategies for a Developing Country," *Food and Nutrition Bulletin* 13, no. 4 (1991): 311–17.

2. Michael Reid, "The Revolution in Retreat—Special Report Cuba," *The Economist*, March 24, 2012; Sara Reardon, "Can Cuban Science Go Global?," *Nature* 537, no. 7622 (2016): 600–603, https://doi.org/ 10.1038/537600a.

3. Vaughan C. Turekian and Norman P. Neureiter, "Science and Diplomacy: The Past as Prologue," *Science & Diplomacy* 1, no. 1 (2012), http://www.science diplomacy.org/editorial/2012/science-and-diplomacy.

4. Sergio Jorge Pastrana and Michael T. Clegg, "US-Cuban Scientific Relations," *Science* 322, no. 5900 (2008): 345, https://doi.org/10.1126/science.1162561.

5. Interview with Manuel Raíces Pérez Castañeda, Cuban Center for Genetic Engineering and Biotechnology, Havana. March 2012.

6. Richard Stone, "Fidel Castro's First-Born Son Foments a Nanotech Revolution," *Science* 348, no. 6236 (2015): 748–49, https://doi.org/ 10.1126 /science.348.6236.748.

7. World Bank Data, GDP (current US$)—Cuba (1970–2020), accessed February 13, 2023, https://data.worldbank.org/indicator/NY.GDP.MKTP. CD?locations=CU.

8. Hugh Thomas, "The Origins of the Cuban Revolution," *The World Today* 19, no. 10 (1963): 448–60, http://www.jstor.org/stable/40393452.

9. David Halberstam, *The Best and the Brightest* (Ballantine, 1992).

10. Anatoly Dobrynin, *In Confidence: Moscow's Ambassador to America's Six Cold War Presidents (1962–1986)* (Times Books, 1995).

11. The White House, "Statement by the President on Cuba Policy Changes," December 17, 2014, https://obamawhitehouse.archives.gov/briefing-room /statements-and-releases.

12. Julie Hirshfield Davis, "US Eases Restrictions on Travel to Cuba and Bank Transactions," *New York Times*, March 15, 2016.

13. The White House, "Statement by the President on Cuba Policy Changes."

14. John R. Pierce and Jim Writer, *Yellow Jack: How Yellow Fever Ravaged America and Walter Reed Discovered Its Deadly Secrets* (Wiley, 2005). Written by the former director of pediatrics at the Walter Reed Army Medical Center, Col. John R. Pierce, this book provides a comprehensive source of documents and letters related to the Yellow Fever Commission, Walter Reed, and Jesse Lazear.

15. Paul Chrystal, *The History of the World in 100 Pandemics, Plagues and Epidemics* (Pen & Sword Books, 2021).

16. Giuseppe Sanarelli, "A Lecture on Yellow Fever. With a Description of the Bacillus Icteroides," *British Medical Journal* 3, no. 2 (1905): 7–11, https://doi .org/10.1136/bmj.2.1905.7.

17. José López Sánchez, *Carlos J. Finlay: His Life and His Work* (Editorial José Martí, 1999).

18. Charles Finlay, trans. Rudolph Matas, "The Mosquito Hypothetically Considered as an Agent in the Transmission of Yellow Fever Poison," *Yale Journal of Biology* 9, no. 6 (1937), 589–604, extracted from *Annals of the Royal Academy of Sciences of Havana,* August 11, 1881.

19. López Sánchez, *Carlos J. Finlay—His Life and His Work.*

20. Dan Cavanaugh, "Walter Reed and the Scourge of Yellow Fever," *UVA Today,* November 13, 2019, https://news.virginia.edu/content/walter-reed-and -scourge-yellow-fever.

21. John R. Pierce, interview and email communications, 2022.

22. Colin Norman, "The Unsung Hero of Yellow Fever?," *Science* 223, no. 4683 (1984): 1370–1372, htttps://doi.org/10.1126/science.6142529.

23. *Philadelphia Medical Journal* 6 (1900): 790–96, reprinted in *Review of Infectious Diseases* 5, no. 6 (1983): 1103–11, https://www.jstor.org/stable /4453230.

24. "The Mosquito Hypothesis," *Washington Post,* November 2, 1900.

25. Akhil Mehra, "Politics of Participation: Walter Reed's Yellow Fever Experi- ments," *AMA Journal of Ethics,* April 2009, https://doi.org/10.1001/virtual mentor.2009.11.4.mhst1-0904.

26. Nobel Prize Nomination Archive, accessed January 29, 2025, https://www .nobelprize.org/nomination/archive/.

27. Juan A. Del Regato, "Carlos Juan Finlay (1833–1915)," *Journal of Public Health Policy* 22, no. 1 (2001): 98–104, https://doi.org/10.2307/3343556.

28. Stephen Gibbs, "Cuba's Castro Recovering Says Son," *BBC News,* February 16, 2007.

29. David Owen, *In Sickness and In Power: Illness in Heads of Government During the Last 100 Years* (Praeger, 2008).

30. Alan Robock, "What Fidel Castro Ruz said in our meeting on December 15 (5:00–8:30 p.m.)" (unpublished personal communication).

31. Gardiner Harris and Adam Goldman, "Illnesses at US Embassy in Havana Prompt Evacuation of More Diplomats," *New York Times,* September 29, 2017;

Carl A. Zimmer, "A 'Sonic Attack' on Diplomats in Cuba? These Scientists Doubt It," *New York Times*, October 5, 2017.

32. Jack Hitt, "The Real Story Behind the Havana Embassy Mystery," *Vanity Fair*, January 6, 2019, *An Assessment of Illness in US Government Employees and Their Families at Overseas Embassies*, ed. David A. Relman and Julie A. Pavlin (Academies Press, 2020).

33. Allan H. Frey, "Human Auditory System Response to Modulated Electromagnetic Energy," *Journal of Applied Physiology* 17 (1962): 689–92.

34. Quoted in Shane Harris and Missy Ryan, "CIA Finds No 'Worldwide Campaign' by Any Foreign Power Behind Mysterious 'Havana Syndrome'," *Washington Post,* January 20, 2022.

35. Conversation with Luis Alberto Montero Cabrera, former president of the Cuban Chemical Society; Richard Stone, "Fidel Castro's First-Born Son Foments a Nanotech Revolution," *Science* 348, no. 6326 (2015): 748–749, https:// doi.org/ 10.1126/science.348.6236.748; Anthony Faiola, "Fidel Castro's Son, a Bookish Nuclear Scientist, Commits Suicide," *Washington Post,* February 1, 2018.

CHAPTER 2. THE ISLAMIC REPUBLIC OF IRAN

In addition to the sources listed below, we have benefited from conversations with Vali Nasr, former dean of Johns Hopkins School for Advanced International Studies (SAIS); Norman Neureiter, former science advisor to US Secretary of State Colin Powell; Thomas R. Pickering, former US Ambassador to the United Nations, Russia, India, Israel, El Salvador, Jordan, and Nigeria; Glenn E. Schweitzer, director of the Office for Central Europe and Eurasia, US National Academies; and Hassan Vafai, former professor of engineering, Sharif University of Technology, Tehran; currently research professor, University of Arizona.

1. Nuclear Threat Initiative, "Iran: Fordow Fuel Enrichment Plant," last updated June 9, 2023, https://www.nti.org/education-center/facilities/fordow -fuel-enrichment-plant/.

2. Associated Press, "Iran's Nuclear Facility Deep Inside a Mountain," November 5, 2019, https://apnews.com/article/middle-east-religion-iran -mountains-iran-nuclear-be54b7a3e8cb4ad788b69b6cf1178109.

3. World Nuclear Association, "Uranium Enrichment," updated November 19, 2024, https://world-nuclear.org/information-library/nuclear -fuel-cycle/conversion-enrichment-and-fabrication/uranium-enrichment .aspx.

4. US Department of Energy, Office of Scientific and Technical Information, "Manhattan Project: Early Uranium Research, 1939–1941," accessed January 29, 2025, https://www.osti.gov/opennet/manhattan-project-history /Events/1939-1942/uranium_research.htm.

5. "Total Casualties: The Atomic Bombings of Hiroshima and Nagasaki," *AtomicArchive.com,* accessed January 29, 2025, https://www.atomicarchive .com/resources/documents/med/med_chp10.html.

6. David E. Sanger and William J. Broad, "US and Allies Warn Iran over Nuclear 'Deception,'" *New York Times,* September 25, 2009.

7. Nuclear Threat Initiative, "Iran: Natanz Enrichment Complex," updated June 13, 2023, https://www.nti.org/education-center/facilities/natanz -enrichment-complex/.

8. "Human Chain Around Fordow Plant," *IranWire,* October 22, 2013, https:// iranwire.com/en/blogs/62027/.

9. Kelsey Davenport, "Iran, P5 + 1 Sign Nuclear Agreement," *Arms Control Association,* December 4, 2013, https://www.armscontrol.org/act/2013-12/iran -p51-sign-nuclear-agreement.

10. Al Jazeera, "UAE's Enoc Pays Iran $4 Billion in Oil Dues," May 29, 2017, https://www.aljazeera.com/economy/2017/5/29/uaes-enoc-pays-iran-4-billion -in-oil-dues.

11. The White House, "The Iran Nuclear Deal: What You Need to Know About the JCBOA," July 14, 2015, https://obamawhitehouse.archives.gov/sites /default/files/docs/jcpoa_what_you_need_to_know.pdf.

12. United States Institute of Peace, "The Iran Primer: Geneva Deal I: Final Terms and Fact Sheet," November 24, 2013, https://iranprimer.usip.org /blog/2013/nov/24/geneva-deal-i-final-terms-and-fact-sheet.

13. The White House, "The Historic Deal that Will Prevent Iran from Acquiring a Nuclear Weapon," accessed January 29, 2025, https://obamawhitehouse .archives.gov/issues/foreign-policy/iran-deal.

14. Ariel Levite, "Can a Credible Nuclear Breakout Time with Iran Be Restored?," *Carnegie Endowment for International Peace,* June 24, 2021, https://carnegieen dowment.org/2021/06/24/can-credible-nuclear-breakout-time-with-iran -be-restored-pub-84833.

15. Steve Inskeep, "Born in the USA: How America Created Iran's Nuclear Program," *NPR,* September 18, 2015, https://www.npr.org/sections /parallels/2015/09/18/440567960/born-in-the-u-s-a-how-america-created -irans-nuclear-program.

16. Ariana Rowberry, "Sixty Years of 'Atoms for Peace' and Iran's Nuclear Program," *Brookings Institution,* December 18, 2013, https://www.brookings .edu/blog/up-front/2013/12/18/sixty-years-of-atoms-for-peace-and-irans -nuclear-program/.

17. Mara Drogan, "The Atoms for Peace Program and the Third World," *Cahiers du Monde Russe* 60, nos. 2–3 (2019): 441–60, https://doi.org/10.4000 /monderusse.11249.

18. Drogan, "The Atoms for Peace Program."

19. "Truman Hints to Stalin About a 'New Weapon of Unusual Destructive Force,'" *History.com,* accessed January 29, 2025, https://www.history.com/this -day-in-history/truman-drops-hint-to-stalin-about-a-terrible-new-weapon.

20. "Japan's Surrender Made Public," *History.com,* accessed January 29, 2025, https://www.history.com/this-day-in-history/japans-surrender-made-public.

21. Jeremy Kuzmarov and Roger Peace, "Was There a Diplomatic Alternative? The Atomic Bombing and Japan's Surrender," *Asia-Pacific Journal,* October 15, 2021, https://apjjf.org/2021/20/kuzmarov-peace.

22. UPI, "Nixon Says He Considered Using Atomic Weapons on 4 Occasions," *New York Times,* July 22, 1985.

23. Raphael Ofek, "Iran's Nuclear Program: Where Is It Going?," *The Begin-Sadat Center for Strategic Studies,* September 5, 2021, https://besacenter.org/wp -content/uploads/2021/09/198web-1.pdf.

24. Lawrence Wu and Michelle Lanz, "How The CIA Overthrew Iran's Democracy In 4 Days," *NPR,* February 7, 2019, https://www.npr.org /2019/01/31/690363402/how-the-cia-overthrew-irans-democracy-in-four -days.

25. "Anglo-Iranian Oil Company," *New World Encyclopedia,* accessed January 29, 2025, https://www.newworldencyclopedia.org/entry/Anglo-Iranian_Oil _Company.

26. Steven Marsh, "The United States, Iran and Operation 'Ajax': Inverting Interpretive Orthodoxy," *Middle East Studies* 39, no. 3 (2003): 1–38, https:// www.jstor.org/stable/4284305.

27. "Mohsen Fakhrizadeh: Iran Blames Israel for Killing Top Scientist," *BBC,* November 28, 2020, https://www.bbc.com/news/world-middle-east -55111064.

28. Quoted in Richard Stone, "Iran's Atomic Czar Explains How He Helped Seal the Iran Nuclear Agreement," *Science,* August 12, 2015.

29. The White House, "The Historic Deal That Will Prevent Iran from Acquiring a Nuclear Weapon," accessed January 29, 2025, https://obamawhitehouse .archives.gov/issues/foreign-policy/iran-deal.

30. The White House, "Statement by the President on Iran," July 14, 2015, https://obamawhitehouse.archives.gov/the-press-office/2015/07/14/statement -president-iran.

31. "Iran Nuclear Deal: What It All Means," *BBC News,* November 23, 2021, https://www.bbc.com/news/world-middle-east-33521655.

32. Brian Murphy, "Iran Claims $100 Billion Now Freed in Major Step as Sanctions Roll Back," *The Washington Post,* February 1, 2016.

33. Jeffrey Lewis, "How Trump Got Suckered by Iran and North Korea," *Vox,* January 8, 2020, https://www.vox.com/world/2020/1/8/21057011/trump -iran-speech-response-foreign-policy.

34. United States Institute of Peace, "Iran's Breaches of the Nuclear Deal," *The Iran Primer,* October 2, 2019, https://iranprimer.usip.org/blog/2019/oct/02 /iran%E2%80%99s-breaches-nuclear-deal.

35. Arms Control Association, "Iran Announces New Nuclear Deal Breach," January 9, 2020, https://www.armscontrol.org/blog/2020-01-09/p4-1-iran -nuclear-deal-alert.

36. Arms Control Association, "Assessing the Risk Posed by Iran's Violations of the Nuclear Deal," *Issue Briefs* 11, no. 9 (December 17, 2019), https://www .armscontrol.org/issue-briefs/2019-12/assessing-risk-posed-iran-violations -nuclear-deal.

37. US Department of the Treasury, Office of Foreign Assets Control, "Ukraine -/Russia-Related Sanctions," accessed January 29, 2025, https://home.treasury .gov/policy-issues/financial-sanctions/sanctions-programs-and-country -information/ukraine-russia-related-sanctions.

CHAPTER 3. THE DEMOCRATIC PEOPLE'S REPUBLIC OF KOREA

1. Cameron Forbes, *The Korean War: Australia in the Giants' Playground* (New York: Macmillan, 2010).

2. Anthony Kuhn, "Hundreds of Thousands of Landmines Remain from Korean War but Serve No Purpose," *NPR,* August 27, 2019, https://www.npr.org.

3. Allan Reed Millett, ed., *The Korean War,* 3 vols. (University of Nebraska Press, 1997–2002).

4. Ralph C. Hassig and Kong Dan Oh, *The Hidden People of North Korea: Everyday Life in the Hermit Kingdom* (Rowman & Littlefield, 2009).

5. Barbara Crossette, "Korean Famine Toll: More than 2 Million," *New York Times,* August 20, 1999.

6. Stuart Thorson and Hyunjin Seo, "Building Partners Through Academic Science," *Asian Perspective* 38, no. 1 (2014): 137–61, http://www.jstor.org/stable/42704857.

7. George S. Bain, "One Korea," December 23, 2009, Maxwell School of Citizenship and Public Affairs, Syracuse University, https://www.maxwell.syr.edu/news/article/one-korea.

8. Kelsey Davenport, "The Six-Party Talks at a Glance," last reviewed February 2023, Arms Control Association, https://www.armscontrol.org/factsheets/6partytalks.

9. Junghoon Ki, Minki Sung, and Choongik Choi, "Impact of Nuclear Tests on Deforestation in North Korea Using Google Earth–Based Spatial Images," *Journal of People, Plants, and Environment* 22, no. 6 (2019): 563–73, https://doi.org/10.11628/ksppe.2019.22.6.563.

10. Grace Lee, "The Political Philosophy of Juche," *Stanford Journal of East Asian Affairs* 3, no. 1 (2003): 105–12.

11. Joseph S. Nye, "Soft Power," *Foreign Policy* 80 (1990): 153–71, https://doi.org/10.2307/1148580.

12. Semoon Chang, "The Saga of US Economic Sanctions Against North Korea," *The Journal of East Asian Affairs* 20, no. 2 (2006): 109–39, https://www.jstor.org/stable/23257941.

13. Ki, Sung, and Choi, "Impact of Nuclear Tests on Deforestation in North Korea."

14. "North Korea Fires Seven Short-Range Missiles into East Sea," *Hankyoreh*, July 6, 2009, https://english.hani.co.kr/arti/english_edition/e_northkorea/364175.html.

15. Hugo Martin, "North Korea's Air Koryo Named the Worst Airline in the World Again," *Los Angeles Times*, August 13, 2016, https://www.latimes.com/business/la-fi-travel-briefcase-koryo-20160813-snap-story.html.

16. James E. Hoare, *Historical Dictionary of Democratic People's Republic of Korea* (Rowman & Littlefield, 2019).

17. UN World Food Programme, *Democratic People's Republic of Korea*, accessed January 30, 2025, https://www.wfp.org/countries/democratic-peoples-republic-korea.

18. James Belgrave [World Food Programme], "Inside the Democratic People's Republic of Korea," May 30, 2019, https://www.wfp.org/stories/inside -democratic-peoples-republic-korea.

19. Council on Foreign Relations, "North Korea's Military Capabilities," last updated June 28, 2022, https://www.cfr.org/backgrounder/north-korea -nuclear-weapons-missile-tests-military-capabilities.

20. Janis Mackey Frayer, "Pyongyang University of Science and Technology: 2 US Employees Detained," *NBCNews.com,* May 8, 2017, https://www.nbcnews.com /news/north-korea/pyongyang-university-science-technology-2-u-s-employees -detained-n756176.

21. Interview with Seema Yasmin.

22. Bruce Cumings, *Korea's Place in the Sun: A Modern History* (W. W. Norton, 1997).

23. Interview with Seema Yasmin.

24. Interview with Seema Yasmin.

25. Christopher Small and Terry Naumann, "The Global Distribution of Human Population and Recent Volcanism," *Global Environmental Change Part B: Environmental Hazards* 3, nos. 3–4 (2001): 93–109, https://doi.org/10.1016 /S1464-2867(02)00002-5.

26. Nsikan Akpan, "Western Scientists Dissect a North Korea Volcano Cut Off by Diplomatic Sanctions," *PBS,* April 15, 2016, https://www.pbs.org/newshour /science/western-scientists-dissect-a-north-korea-volcano-cutoff-by-diplomatic -sanctions.

27. Maddalena Ragona, Francesca Hannstein, and Mario Mazzocchi, "The Financial Impact of the Volcanic Ash Crisis on the European Airline Industry," *CESifo Forum* 11, no. 2 (2010): 92–100, https://ideas.repec.org/a/ces/ifofor /v11y2010i02p92-100.htm.

28. Richard B. Stothers, "The Great Tambora Eruption in 1815 and Its Aftermath," *Science* 224, no. 4654 (1984): 1191–98, http://www.jstor.org/stable /1692039.

29. Interview with Seema Yasmin; Richard Stone, "Sizing Up a Slumbering Giant," *Science* 341, no. 6150 (2013): 1060–61, https://doi.org/10.1126/science.341.6150.1060.

30. Interview with Seema Yasmin; Stone, "Sizing Up a Slumbering Giant."

31. Interview with Seema Yasmin; Stone, "Sizing Up a Slumbering Giant."

32. Interview with Seema Yasmin; Stone, "Sizing Up a Slumbering Giant."

33. Interview with Seema Yasmin; Stone, "Sizing Up a Slumbering Giant."

34. Interview with Seema Yasmin; Stone, "Sizing Up a Slumbering Giant."

35. J. Stephen Morrison et al., "The Gathering Health Storm Inside North Korea," Center for Strategic and International Studies, May 7, 2018, https://www.csis.org/analysis/gathering-health-storm-inside-north-korea; Richard Stone and Cathleen A. Campbell, "A New Era of Forging Connections and Trust with North Korea's Scientists," *Science & Diplomacy*, October 4, 2018, https://www.sciencediplomacy.org/perspective/2018/new-era-forging-connections-and-trust-north-koreas-scientists.

36. Peter H. Raven, "Engaging North Korea through Biodiversity Protection," *Science & Diplomacy*, September 9, 2013, https://www.sciencediplomacy.org/perspective/2013/engaging-north-korea-through-biodiversity-protection.

37. S. Y. Kim, S. Y. Park, and K. S. Park, "A Study on New Change of Forest Management in DPRK by Introducing Agroforestry of Sloping Land Management," *The Korean Journal of Unification Affairs* 28, no. 2 (2016): 127–57.

38. Jeong Seo-Yeong, "Sources: President of State Academy of Sciences Dismissed," *Daily NK*, December 31, 2019, https://www.dailynk.com/english/sources-president-of-state-academy-of-sciences-dismissed/; Joseph Hincks, "'Worse Than Nazi Camps': New Report Details Gruesome Crimes Against Humanity at North Korean Prisons," *Time*, December 12, 2017, https://time.com/5060144/north-korea-political-prisons/.

39. Krishnadev Calamur, "The Cost of Stealing a Sign: 15 Years of Hard Labor," *The Atlantic*, March 16, 2016, https://www.theatlantic.com/international/archive/2016/03/north-korea-american-tourist-sentenced/474000/.

40. "Factbox: The Three Americans Imprisoned by North Korea, *Reuters*, March 12, 2018, https://www.reuters.com/article/northkorea-missiles-detainees -idINKCN1GO0A9.

41. Junichi Fukuda, "Next Phase of North Korean Missile Tests: A New ICBM and Other Developments, January–April 2022," *International Information Network Analysis* (Sasakawa Peace Foundation), June 16, 2022, https://www.spf.org /iina/en/articles/fukuda_03.html.

CHAPTER 4. SUB-SAHARAN AFRICA

1. Centers for Disease Control and Prevention, *Laboratory Identification of Parasites of Public Health Concern: Malaria,* last reviewed September 24, 2024, https://www.cdc.gov/dpdx/malaria/index.html.

2. Ker Than, "King Tut Mysteries Solved: Was Disabled, Malarial, and Inbred," *National Geographic News*, February 17, 2010.

3. Brian Greenwood, "Between Hope and a Hard Place," *Nature* 430 (2004): 926–927, https://doi.org/10.1038/430926a.

4. Apoorva Mandavilli, "A 'Historic Event': First Malaria Vaccine Approved by W.H.O.," *New York Times*, October 6, 2021.

5. World Health Organization, *World Malaria Report 2022*, December 8, 2022, https://www.who.int/teams/global-malaria-programme/reports/world-malaria -report-2022.

6. Centers for Disease Control and Prevent and Association of Public Health Laboratories, *Hemoglobinopathies: Current Practices for Screening, Confirmation and Follow-up*, December 2015, https://www.cdc.gov/sickle-cell/media/pdfs /nbs_hemoglobinopathy-testing_122015.pdf.

7. Carl E. Taylor, Jeanne S. Newman, and Narindar U. Kelly, "The Child Survival Hypothesis," *Population Studies* 30, no. 2 (1976): 263–78, https://doi.org /10.2307/2173609.

8. Centers for Disease Control and Prevention, *CDC's Origins and Malaria*, https://www.cdc.gov/museum/history/our-story.html.

9. Joel C. Brennan, Andréa Egan, and Gerald T. Keusch, *The Intolerable Burden of Malaria: A New Look at the Numbers* (supplement to vol. 64, no. 1, of *The American Journal of Tropical Medicine and Hygiene*) (American Society of Tropical Medicine and Hygiene, 2001), https://www.ncbi.nlm.nih.gov/books/NBK2622/.

10. Nayantara Sarma, Edith Patouillard, Richard E. Cibulskis, and Jean-Louis Arcand, "The Economic Burden of Malaria: Revisiting the Evidence," *American Journal of Tropical Medicine and Hygiene* 101, no. 6 (2019): 1405–15, https://doi.org/10.4269/ajtmh.19-0386.

11. Kenneth J. Arrow, Claire Panosian, and Hellen Gelband, eds., *Saving Lives, Buying Time: Economics of Malaria Drugs in an Age of Resistance* (National Academies Press, 2004).

12. *William Crawford Gorgas* (1854–1920), https://encyclopediaofalabama.org/article/william-crawford-gorgas/.

13. European Centre for Disease Prevention and Control, *Aedes aegypti: Factsheet for Experts*, last updated January 2, 2023, https://www.ecdc.europa.eu/en/disease-vectors/facts/mosquito-factsheets/aedes-aegypti.

14. David Martínez, Carolina Hernández, Marina Muñoz, Yulieth Armesto, Andres Cuervo, and Juan David Ramírez, "Identification of *Aedes* (Diptera: Culicidae) Species and Arboviruses Circulating in Arauca, Eastern Colombia," *Frontiers in Ecology and Evolution* 8 (2020), https://doi.org/10.3389/fevo.2020.602190.

15. Malcolm Gladwell, "The Mosquito Killer," *The New Yorker*, June 24, 2001.

16. Perrine Marcenac, W. Robert Shaw, Evdoxia G. Kakani, Sara N. Mitchell, Adam South, Kristine Werling . . . Flaminia Catteruccia, "A Mating-Induced Reproductive Gene Promotes Anopheles Tolerance to Plasmodium Falciparum Infection," *PLOS Pathogens* 16, no. 12 (2020): e1008908, https://doi.org/10.1371/journal.ppat.1008908.

17. National Library of Medicine, Profiles in Science, Fred L. Soper Papers, *Fighting Yellow Fever and Malaria in Brazil, 1928–1942*, accessed January 30, 2025, https://profiles.nlm.nih.gov/spotlight/vv/feature/campaign.

18. Jake Smith, "A Fool's Errand? Eliminating Mosquitoes to End Disease Epidemics," *WHYY,* June 9, 2016, https://whyy.org/segments/a-fools-errand -eliminating-mosquitos-to-end-disease-epidemics/; National Library of Medicine, Profiles in Science, Fred L. Soper Papers, *Biographical Overview,* accessed January 30, 2025, ttps://profiles.nlm.nih.gov/spotlight/vv/feature /biographical-overview.

19. Pan American Health Organization, *History of the Pan American Health Organization,* accessed January 30, 2025, https://www.paho.org/en/who-we -are/history-paho.

20. Linda Lear, "Rachel Louise Carson: Biography," revised 2015, *The Life and Legacy of Rachel Carson,* https://www.rachelcarson.org/biography.

21. Carolyn Kimmel, "Two Dillsburg Doctors Bring Help and Hope to a Village in Africa," *Patriot-News,* August 25, 2013, https://www.pennlive.com/bodyandmind /2013/08/two_harrisburg-area_doctors_br.html.

22. Brethren in Christ Historical Society, *Sikalongo Mission 100th Anniversary*, September 9, 2016, https://bic-history.org/photo-friday-sikalongo-mission -100th-anniversary/.

23. Macha Research Trust, *What We Do,* accessed January 30, 2025, https://www .macharesearch.org/what-we-do.

24. Mat Edelson, "Mission Man," *Hopkins Bloomberg School of Public Health Magazine,* January 11, 2011, https://magazine.publichealth.jhu.edu/2011 /mission-man.

25. *John Hopkins Malaria Research Institute*, accessed January 30, 2025, https:// malaria.jhsph.edu/malaria-institute-at-macha-miam/macha-area/.

26. *Malaria Diagnosis (United States).* (n.d.). https://www.cdc.gov/malaria /diagnosis_treatment/diagnosis.html.

27. P. E. Thuma, G. J. Bhat, G. F. Mabeza, C. Osborne, G. Biemba, G. M. Shakankale, P. A. Peeters, B. Oosterhuis, C. B. Lugt, and V. R. Gordeuk, "A Randomized Controlled Trial of Artemotil (Beta-arteether) in Zambian Children with Cerebral Malaria," *American Journal of Tropical Medicine and*

Hygiene 64, no. 4 (2000): 524–29, https://doi.org/10.4269/ajtmh.2000 .62.524.

28. Novartis News Release, "Novartis Reaches Milestone Delivery of 1 Billion Courses of Antimalarial Treatment," May 18, 2021, https://www.novartis.com /news/media-releases/novartis-reaches-milestone-delivery-1-billion-courses -antimalarial-treatment.

29. Joel G. Breman, Martin S. Alilio, and Nicholas J. White, "Defining and Defeating the Intolerable Burden of Malaria, III: Progress and Perspectives" (supplement to vol. 77, no. 6 [2007] of the *American Journal of Tropical Medicine and Hygiene*), https://www.ajtmh.org/view/journals/tpmd/77/6_Suppl /article-pvi.pdf.

30. Sungano Mharakurwa, Mavis Sialumano, Kun Liu, Alan Scott, and Philip Thuma, "Selection for Chloroquine-Sensitive *Plasmodium Falciparum* by Wild *Anopheles arabiensis* in Southern Zambia," *Malaria Journal* 12 (2013): article 453, https://doi.org/10.1186/1475-2875-12-453.

31. "Zambia: Malaria Death Rate Plummets Through Boost in Control Measures," *UN News,* April 23, 2009, https://news.un.org/en/story/2009/04/297772.

32. Cathy Shufro, "All Malaria Is Local," *Hopkins Bloomberg School of Public Health Magazine,* April 13, 2022, https://magazine.jhsph.edu/2022/all-malaria -local.

33. Victor M. Mukonka, Emmanuel Chanda, Ubydul Haque, Mulakwa Kamuliwo, Gabriel Mushinge, Jackson Chileshe, Kennedy A. Chibwe, Douglas E. Norris, Modest Mulenga, Mike Chaponda, Mbanga Muleba, Gregory E. Glass, and William J. Moss, "High Burden of Malaria Following Scale-Up of Control Interventions in Nchelenge District, Luapula Province, Zambia," *Malaria Journal* 13 (2014): article 153, https://doi.org/10.1186/1475-2875-13-153.

34. Dana al-Hindi and Brenna Henn, "When Does 'Helicopter Research' Turn Exploitative?," *Science: The Wire,* April 20, 2022, https://science.thewire.in /the-sciences/helicopter-research-exploitation-khoe-san-communities-nagoya -dna-ownership/.

35. Farid Dahdouh-Guebas, J. Ahimbisibwe, Rita Van Moll, and Nico Koedam, "Neo-Colonial Science by the Most Industrialised upon the Least Developed Countries in Peer-Reviewed Publishing," *Scientometrics* 56, no. 3 (2003), https://doi.org/10.1023/A:1022374703178.

36. The European & Developing Countries Clinical Trials Partnership, *Prof. Godfrey Biemba*, accessed January 30, 2025, http://www.edctp.org/about-us /governance/board-edctp-association/members-of-the-board/prof-godfrey -biemba/.

37. The Special Programme for Research & Training in Tropical Diseases: *Dr. Modest Mulenga*, accessed January 30, 2025, https://profiles.tdr-global.net /Modest.Mulenga.

38. Macha Research Trust, *Annual Report 2011*.

39. Africa University, *Academic Deans*, accessed January 30, 2025, https://www .africau.edu/deans.html.

40. Georgia Department of Public Health, *Lynn Paxton, M.D., M.P.H*, accessed January 30, 2025, https://dph.georgia.gov/lynn-paxton-md-mph.

41. US Department of State, Office of International Religious Freedom, *2021 Report on International Religious Freedom: Tanzania*, accessed January 30, 2025, https://www.state.gov/reports/2021-report-on-international-religious-freedom /tanzania/.

42. UN Refugee Agency, Global Law & Policy Database, *World Directory of Minorities and Indigenous Peoples—United Republic of Tanzania: Shirazi and Arabs of Zanzibar*, 2008, https://www.refworld.org/reference/countryrep/mrgi /2008/en/64031.

43. Kaiser Family Foundation, *The President's Malaria Initiative and Other US Government Global Malaria Efforts*, December 20, 2023, https://www.kff.org /global-health-policy/.

44. The Global Fund, *PEPFAR and The Global Fund Collaborate to Treat 3.7 Million Living with HIV/AIDS*, December 1, 2009, https://www.theglobalfund.org/en /news/2009/.

45. The White House, *U.S.-Africa Partnership in Health Cooperation*, December 13, 2022, https://www.whitehouse.gov/briefing-room/statements-releases /2022/12/13/fact-sheet-u-s-africa-partnership-in-health-cooperation/.

46. Kaiser Family Foundation, *PEPFAR Reauthorization: Side-by-Side of Legislation Over Time*, August 18, 2022, https://www.kff.org/global-health-policy/.

47. US Department of State, *Reimagining PEPFAR's Strategic Direction Fulfilling America's Promise to End the HIV/AIDS Pandemic by 2030*, September 2022, https://www.state.gov/wp-content/uploads/2022/09/PEPFAR-Strategic -Direction_FINAL.pdf.

48. Kaiser Family Foundation, *The US President's Emergency Plan for AIDS Relief (PEPFAR)*, July 12, 2022, https://www.kff.org/global-health-policy/.

49. "Anti-Prostitution Pledge in US Aids Funding 'Damaging' HIV Response," *The Guardian,* July 24, 2012, https://www.theguardian.com/global-development /2012/jul/24/prostitution-us-aids-funding-sex.

50. Fidel Zavala, "RTS,S: The First Malaria Vaccine," *Journal of Clinical Investigation* 132, no. 1 (2022): e156588, https://doi.org/10.1172/JCI156588; World Health Organization, "WHO Recommends R21/Matrix-M Vaccine for Malaria," Press release, October 2, 2023, https://www.who.int/news.

CHAPTER 5. SCIENCE ON TRIAL

1. The Nobel Prize, "Peter Agre and Roderick MacKinnon. The Nobel Prize in Chemistry 2003," accessed January 30, 2025, https://www.nobelprize.org /prizes/chemistry/2003/summary/.

2. National Academies of Sciences, Engineering, and Medicine, "About the Committee on Human Rights," accessed January 30, 2025, https://www .nationalacademies.org/chr/about.

3. The Andrei Sakharov Science Endowment Fund, "Andrei Sakharov: Biography," accessed January 30, 2025, https://sakharov.fund/en/andrei -sakharov/bio/.

4. Jay Bergman, *Meeting the Demands of Reason: The Life and Thought of Andrei Sakharov* (Cornell University Press, 2009).

5. New World Encyclopedia, "Dr. Andrei Sakharov," accessed January 30, 2025, https://www.newworldencyclopedia.org/entry/Andrei_Sakharov.

6. World Summit of Nobel Peace Laureates, "Andrei Sakharov," accessed January 30, 2025, http://www.nobelpeacesummit.com/project/andrei -sakharov/.

7. The Nobel Prize, "Andrei Sakharov: Biographical," accessed January 30, 2025, https://www.nobelprize.org/prizes/peace/1975/sakharov/biographical/.

8. The Nobel Prize, "Andrei Sakharov: Acceptance Speech," December 10, 1975, https://www.nobelprize.org/prizes/peace/1975/sakharov/acceptance -speech/.

9. "Sakharov Exile Ends; He'll Return to Post in Moscow," *Los Angeles Times*, December 19, 1986, https://www.latimes.com/archives/la-xpm-1986-12-19-mn -3708-story.html.

10. International League for Human Rights, *Andrei Sakharov from Exile*, October 1983, https://history.aip.org/exhibits/sakharov/from-exile.html.

11. Committee on Human Rights, National Academies of Sciences, Engineering, and Medicine, *Advocacy for Health Professionals: Regional Breakdown of Cases Concerning Health Professionals*, 2021.

12. Committee on Human Rights, National Academies of Sciences, Engineering, and Medicine, *Confronting Human Rights Abuse: A Guide for Supporting Scientists, Engineers, and Health Professionals Under Threat* (2021), https://www .confront-rights-abuse.org/.

13. Committee on Human Rights, National Academies of Sciences, Engineering, and Medicine, *Advocacy*, accessed January 30, 2025, https://www.national academies.org/chr/advocacy.

14. Reuters, "Profiles of Foreign Medics in Libyan HIV Case," August 9, 2007, https://www.reuters.com/article/us-libya-trial-medics-factbox/factbox -profiles-of-foreign-medics-in-libyan-hiv-case-idUSL186433220061219.

15. Associated Press, "The Bulgarian Medics, the Libyan Children and the HIV Epidemic," *The Guardian*, July 24, 2007, https://www.theguardian.com/world /2007/jul/24/libya1.

16. "A Shocking Lack of Evidence," *Nature* 443, no. 7114 (2006): 888–89, https://doi.org/10.1038/443888a.

17. Declan Butler, "Europe Condemns Libyan Trial Verdict," *Nature* 445, no. 7123 (2007): 7, https://doi.org/10.1038/445007a.

18. "Profiles: The Imprisoned Medics," *BBC News*, July 24, 2007, http://news.bbc.co.uk/2/hi/africa/6896231.stm.

19. Anna Mudeva, "Palestinian Doctor Will Not Forgive Libyan Jailors," *Reuters*, August 9, 2007, https://www.reuters.com/article/us-libya-nurses-palestinian/palestinian-doctor-will-not-forgive-libyan-jailors-idUSL2687636620070726.

20. Human Rights Watch, "Libya: Foreign Health Workers Describe Torture," November 13, 2005, http://www.hrw.org/news/2005/11/13/libya-foreign-health-workers-describe-torture.

21. Declan Butler, "Lawyers Call for Science to Clear AIDS Nurses in Libya," *Nature* 443, no. 7109 (2006): 254–55, https://doi.org/10.1038/443254b.

22. Declan Butler, "Molecular HIV Evidence Backs Accused Medics," *Nature* 444, no. 7120 (2006): 658–59, https://doi.org/10.1038/444658b.

23. Vittorio Colizzi, Tulio de Oliveira, and Richard J. Roberts, "Libya Should Stop Denying Scientific Evidence on HIV," *Nature* 448, no. 7157 (2007): 992, https://doi.org/10.1038/448992a.

24. John Biewen and Ian Ferguson, "Shadow over Lockerbie," *American Public Media*, March 2000, https://americanradioworks.publicradio.org/features/lockerbie/story/printable_story.html.

25. John Bohannon, "Pressure Mounts on Libya to Free Medical Workers," *Science*, October 24, 2006, https://www.science.org/content/article/pressure-mounts-libya-free-medical-workers; Elizabeth Rosenthal, "HIV Trial in Libya Is Criticized," *New York Times*, November 5, 2006.

26. R. D. Putnam, "With Libya's Megalomaniac 'Philosopher-King'," *Wall Street Journal*, February 26, 2011.

27. "Profile: Abdullah al-Senussi," *BBC News*, October 26, 2015, https://www.bbc.com/news/world-middle-east-17414121; Ian Black, "Abdullah al-Senussi: Spy

Chief Who Knew Muammar Gaddafi's Secrets," *The Guardian*, September 5, 2012, https://www.theguardian.com/.

28. Jeevan Vasagar, "Academic Linked to Gaddafi's Fugitive Son Leaves LSE," *The Guardian*, October 31, 2011.

29. Robert Putnam, "With Libya's Megalomaniac 'Philosopher King'," *Wall Street Journal*, February 26, 2011.

30. Richard J. Roberts and 113 Nobel Laureates, "An Open Letter to Colonel Muammar al-Gaddafi," *Nature* 444, no. 7116 (2011): 146, https://doi.org/10.1038/444146a; Declan Butler, "High Noon in Libya," *Nature* 448, no. 7151 (2007): 230–31, https://doi.org/10.1038/448230a.

31. Richard J. Roberts and 113 Nobel Laureates, "An Open Letter to Colonel Muammar al-Gaddafi," *Nature* 444, no. 146 (2006): 146, https://doi.org/10.1038/444146a.

32. Declan Butler, "Diplomatic Talks Spur Hope in Libya HIV Case," *Nature* 447, no. 7145 (2007): 624, 625, https://doi.org/10.1038/447624b.

33. Reuters, "EU Agrees Care for Libya HIV Children," June 22, 2007, https://www.reuters.com/article/idUKL2286699620070622.

34. Matthew Brunwasser, "Workers Arrive in Bulgaria After Being Freed," *New York Times*, July 24, 2007.

35. Declan Butler, "Libyan Ordeal Ends: Medics Freed," *Nature* 448, no. 7152 (2007): 398, https://doi.org/10.1038/448398a.

36. "Gaddafi's Son to Run for President of Libya," *Al Jazeera*, November 14, 2021, https://www.aljazeera.com/news/2021/11/14/son-of-former-libyan-leader-gaddafi-runs-for-president-official.

37. Scott Neuman, "Former French President Nicolas Sarkozy Found Guilty of Corruption," *NPR*, March 1, 2021, https://www.npr.org/2021/03/01/972453743/former-french-president-sarkozy-found-guilty-of-corruption-sentenced-to-jail.

EPILOGUE

1. "Poll: Arabs Less Favourable to US," *Al Jazeera*, July 23, 2004, https://www.aljazeera.com/news/2004/7/23/poll-arabs-less-favourable-to-us.

Index

JAMES BELLINGHAM, PhD WITH CLAUDIA GEIB
How Are Marine
Robots Shaping
Our Future?

What will robots discover as they
explore life and resources from
the depths of Earth's oceans to
the reaches of deep space?

ARTURO CASADEVALL, MD, PhD
WITH STEPHANIE DESMON, MA
What If Fungi Win?

Explores the beneficial roles of
fungi, their deadly impacts, and
how experts are researching
treatments to save lives and our
food supplies.

SABINE STANLEY, PhD WITH JOHN WENZ
What's Hidden
Inside Planets?

A guided journey through the
inner workings of Earth, the
cloaked mysteries of other
planets in our solar system,
and beyond.

RAMA CHELLAPPA, PhD WITH ERIC NILLER
Can We Trust AI?

Artificial intelligence is part
of our daily lives. How can we
address its limitations and
guide its use for the benefit of
communities worldwide?

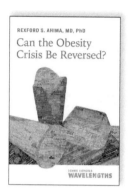

REXFORD S. AHIMA, MD, PhD

Can the Obesity
Crisis Be Reversed?

WAVELENGTHS

How can we work together to
understand the rise of obesity
and reverse its related diseases
and societal impacts?

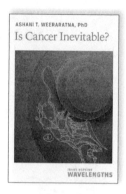

ASHANI T. WEERARATNA, PhD

Is Cancer Inevitable?

WAVELENGTHS

How can new understandings
about cancer cell interactions
help doctors better control, and
eventually cure, cancer?

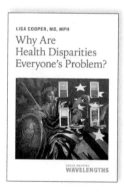

LISA COOPER, MD, MPH

Why Are
Health Disparities
Everyone's Problem?

WAVELENGTHS

How can we all work together to
eliminate the avoidable injustices
that plague our health care
system and society?

JESSICA FANZO, PhD

Can Fixing Dinner
Fix the Planet?

WAVELENGTHS

How can consumers, nations,
and international organizations
work together to improve food
systems before our planet loses
its ability to sustain itself?

JOHNS HOPKINS UNIVERSITY PRESS | PRESS.JHU.EDU |